Dane County Place-Names

DANE COUNTY PLACE-NAMES

by

Frederic G. Cassidy

THE UNIVERSITY OF WISCONSIN PRESS

The University of Wisconsin Press
1930 Monroe Street, 3rd Floor
Madison, Wisconsin 53711-2059
uwpress.wisc.edu

3 Henrietta Street
London WC2E 8LU, England
eurospanbookstore.com

1 3 5 4 2

Printed in the United States of America
Library of Congress Catalog Card Number 68-9024

Much of this book is reproduced from *The Place-Names of Dane County, Wisconsin*, by Frederic G. Cassidy, Publication of the American Dialect Society, Number 7, April 1947.

To Dr. and Mrs. Edwin B. Fred,
whose generous encouragement of this work
came at a time,
twenty years ago, when I would not have
dared to ask for it.

CONTENTS

FOREWORD TO THE 2009 EDITION

David Medaris

Sometime in the early 1990s, an envelope landed on the desk of a columnist for *Isthmus*, the Madison, Wisconsin, weekly. It contained an inquiry regarding the proper name for a flowage extending from Lake Wingra to Lake Monona. The correspondent had heard it called Wingra Creek, but also Murphy's Creek. Which was correct?

Intrigued, the columnist went looking for the answer. The usual resources of the day turned up no definitive leads. Before making the trip down State Street to the Wisconsin Historical Society, he strolled into *Isthmus* editor Marc Eisen's office to ask whether *he* might have any leads. Eisen thought for a few seconds, turned to scan his bookshelves, then reached out, retrieved a tidy little volume and handed it to the columnist.

I was that columnist, and thus began my affection for *Dane County Place-Names*. Frederic G. Cassidy's localized dictionary of geographical etymologies did indeed contain the answer to the question about Murphy's and Wingra creeks. But it was also a vast trove of delights and revelations. Here were hundreds of entries for Dane County's named places, some amounting to narratives as exquisite as they were concise, many dispelling myths or misconceptions, others clarifying a nomenclature with the authority of Cassidy's exhaustive research, some resisting his consultation of maps and historical records but yielding to his capacity for critical thought. I sought a copy of *Dane*

County Place-Names for myself, soon found one at Paul's Book Store, then set about seeking other copies to give to friends. As the years passed, affordable used copies became ever more difficult to find. At some point, the time I spent extolling its many virtues overtook the rate of success at finding yet more copies to give as gifts. You can imagine my excitement at the news that the University of Wisconsin Press would be bringing *Dane County Place-Names* back into print.

Now, I can return to buying copies to give as gifts to friends, while continuing to extol its virtues. And its virtues are many.

To the *x* and *y* axes of the Dane County map, *Dane County Place-Names* adds a *z* axis—a third dimension of historical topography. The time and effort Cassidy invested in researching maps, plat books and other resources—digging down through their documentary layers to mine their ores—continues to pay dividends. The names of many of the places here may have been rendered obvious by their pervasive contemporary use. Cassidy nevertheless provides insights into our most common place-names that may surprise even lifelong Dane County residents. And there are a good many political, social, and natural features whose names Cassidy helps rescue from obscurity and the passage of time, if not extinction—place-names that, while they remain inscribed on nineteenth-century Dane County maps, have fallen out of currency.

Place-names are, after all, as mutable as maps themselves. The ground shifts underfoot. Glaciers reshape the landscape, leaving moraines and kamens to await christening. Rivers and creeks alter course. Catastrophic floods overwhelm a levee and a lake empties. Developers transform farmsteads into new subdivisions. Towns mature into cities and annex other municipalities. Small settlements are abandoned and their buildings toppled. Grand schemes for new plats go unrealized or unfinished. A new donor's generosity overwrites the name of a philanthropist from an earlier generation.

Cassidy's efforts first preserved Dane County's place-names as they had been mapped and recorded up to 1947, when an early version of this work was published by the American Dialect Society. When the University of Wisconsin Press published *Dane County Place-Names* in 1968, the names of presidents, early settlers, developers, prominent local figures, evocative pronouns and adjectives were once again affixed at that point in time to our schools and taverns, townships and villages, post offices and crossings, creeks and marshes, heights and hollows, lakes and landings, rivers and fords, railroads and highways, bluffs and valleys, and dozens of other features we are predisposed to single out with specific names.

Cassidy's appetite for such an undertaking was enormous. He would go on to launch and shepherd the seminal Dictionary of American Regional English project. Even in his nineties, as he was bequeathing DARE's editorial stewardship to his successor, Joan Houston Hall, Cassidy was imbued with the aura of a Great Mind: boundlessly inquisitive, with a taxonomist's aptitude for classification and the eyes of a man blessed with acute vision in both the literal and metaphorical senses. Those qualities are in ample evidence here in this concise volume of little wonders. Perhaps the best insight into the ways he thought and worked can be found in the appendixes, where Cassidy details his methodologies and their underlying rationales, charts and analyzes Dane County's place-names, lists the scores of historical maps and other primary documents he relied on, and cites the names of more than one hundred experts and long-time county residents he consulted—including the inestimable likes of Charles E. Brown and August Derleth.

Cassidy's appendixes make a fine preface to savoring this book cover to cover. *Dane County Place-Names* is, of course, also well-suited to seeking out answers to questions regarding the names of specific places (such as Murphy's Creek) or to turning to any random page and letting your eyes land on any random entry. If your family settled in Dane County during the nineteenth century, or your neighbors include descendants of a long-time Dane County family, you may find the familiar surname in these pages. If not, you may be interested to learn that the Yahara River has been called by a handful of aliases or who christened one of the county's most popular cycling destinations Paoli (and why) or stumble across any one of hundreds of other curiosities contained amid the thousands of points of fact Cassidy has compiled.

But *Dane County Place-Names* is more than a dictionary of local nomenclatures, more than a compendium of historical facts and topographical curiosities. It is the kind of wondrous little volume that compels you to put it down, to get up and seek out the places named, to visit them firsthand, and to understand this place where you live. It is also a book that merits being carried, along with a map. A good map can help you pinpoint your geographical location and find your way, but you need *Dane County Place-Names* to arrive at a deeper sense of where you are.

PREFACE TO THE 1968 EDITION

Dane County, Wisconsin. Why the name "*Dane*"? The question might idly cross the mind of a newcomer. Would we be able to answer it? Or know where to find the answer? Some people seem to think it refers to Scandinavian settlers, but they are wrong. Not a great many Danes have settled in this state—nothing to compare with the large numbers of Norwegians, or with the Swedes, Finns, and even Icelanders. Then why *Dane* County?

The answer will be found in its alphabetical place inside this book—as also the fact that there are within the county a township and a village by the same name, and that before these gained an official existence there had been a *Dane Postoffice* (going back to 1844) which did a deal of moving about before it settled down in the present village. Also a *Dane Precinct* for election purposes (1846). Further, with the coming of the railroad in 1871 a *Dane Station* was established, and following that, *Dane Station Postoffice,* which competed for a time with *Dane Postoffice.*

Who cares about all these dusty old facts? Well, they might be expected to hold some interest for people who call Dane County home. But there is more than that. Dane County is interesting in regard to place-names because, within its borders, one can find virtually all the kinds of names commonly used in Wisconsin and our surrounding states. Dane County, for example, has its share of Indian names—a number genuinely local: *Taychopera, Wingra, Pecatonica;* a good many more artificially grafted in: *Mendota, Monona, Waubesa, Kegonsa, Nakoma;* and a few which look like ingenious inventions: *Winnequah, Tonyawatha.*

There are names from the "Yankee" areas farther east: *Roxbury, Deerfield, Cambridge, Rome,* and names recording the European origins of many settlers: *Westport, Christiana, London.* Famous people have their tribute in the names *Madison, Blackhawk, Ella Wheeler Wilcox, Muir, Vilas;* pioneer settlers too, in *Marxville, Crockertown, Remy Spring, Stoughton, Turville's Point, Spaanem Hill,* and a hundred more. Humorous names are plentiful: *Angleworm Station, Dogtown, Whiskey Creek, Peculiar, Starvation Hill;* also the "hollow" names which amused a former day: *Pumpkin Hollow, Skunk Hollow, Poverty Hollow, Hessian Hollow, Brag Hollow.* Names having a touch of fancy are *Primrose, Mazomanie, Old Pompey;* those expressing the ideals of settlers: *Forward, Liberty Prairie, Hope Corner, Grit Postoffice.* There are Biblical names: *Syene, Mount Horeb;* and descriptive names: *Round Top, Blue Mounds, Sawmill Bottoms, Rattlesnake Bluff, Sugar Bush Point, Yellow Banks, Sun Prairie.* And folklore has attached itself, in defiance of the skeptical, to *Halunkenburg, Pancake Valley, Badfish Creek.*

In short, Dane County offers a harvest of names past and present which should stir a variety of appetites. And if you are curious about the way in which names spread—which feature tends to be named first, and in what pattern, if any, names are transferred from one feature to another—this is followed out in Appendix I.

This study was first published in 1947 (*Publication of*

the American Dialect Society, No. 7, Baltimore) and is reprinted now with the few corrections that proved necessary, but without any further attempt to update it. In these twenty years, of course, changes have taken place; some names have fallen into disuse as small settlements disintegrated, schools were consolidated, roads were relocated. But the changes have been more in the features referred to than in the names themselves. The Town of Madison, for example, has been gradually chewed away by the City of Madison, the latter expanding its corporate area at the expense of the surrounding towns. But this does not affect the name *Madison* as such, its origin, reference, and type. Many new names have been added, chiefly by real estate developers, but they follow established patterns already sufficiently dealt with in the original edition. In other words, the facts displayed here from the first settlement of the County—even before it was formally established—are still valid today. The reader has only to keep this terminal date of 1947 in mind when reading the entries. Virtually all the place-names of present Dane County are listed and explained, as far as our historical information goes.

Now what was that about the source of the name *Dane*?

F. G. C.

Madison, Wisconsin
July 1968

ACKNOWLEDGMENTS

My thanks go to the many informants (see list in Appendix II) who helped me with facts, suggestions, and criticism; to the Wisconsin State Historical Society for the use of its fine library; to Miss Alice E. Smith, curator of MSS. there, for her keen and generous help; to the Graduate School of the University of Wisconsin for financial assistance in doing the field work and clerical work; and to the State Highway Commission of Wisconsin for permission to include their map of Dane County.

INTRODUCTION TO THE 2009 EDITION

Tracy Will

Place-name study offers an understanding of the cultural significance of geographic places through their descriptive or historically applied names. Name origins range from apocryphal native terms to names of influential property owners, from the names of early settlers to the family members of subdivision builders. Frederic G. Cassidy's study *The Place-Names of Dane County, Wisconsin*, published in 1947 and reprinted by the University of Wisconsin Press in 1968, remains a valuable example of the many diverse and interesting studies of geographical areas, whether political or geographical, originating in the 1930s and published after World War II. While many texts in this genre, guided by conventions established by the English Place-Name Society, establish a topical framework to direct their study, Cassidy's *Dane County Place-Names* is guided primarily by the "New English Dictionary" (Oxford). So guided, Cassidy offered a comprehensive review of Dane County's place-names in a facile system for readers to find the origins and explanations for the names of places in Dane County. His dictionary approach may be the most important structural facet of his study, but its descriptive component was also informed by the work of Robert Ramsay in his initial study of Missouri place-names. Ramsay developed various place-name topoi that provided a consistent set of descriptors to be used to illustrate each alphabetically listed place-name. Ramsay's study used the topoi to begin the process of discerning place-name origins, which began with the definition of a place-name descriptor, a narrative that employed a few of the place-names in proving each term's utility as a place-name descriptor, and finally Ramsay listed the places that fell under that definitional rubric. The study is interesting and most reviewers found the narrative portion of greatest interest, but Ramsay's approach is difficult to use for reference or research purposes. Its difficulty arises because the reader must read all the definitional topics first before discovering whether or not the place they are interested in is included on that list.

Rather than follow Ramsay's topoi-driven approach for Dane County, Cassidy followed the "New English Dictionary" model of providing readers with Dane County place-names in an alphabetical dictionary format. Each alphabetic entry is followed by a description of its place-name using several arcana, including all known spellings, original source usage, map origins, common usage, and other factors. Each location includes a definition made up of a series of natural and manmade terms used to describe thousands of Dane County locations. Cassidy listed each entry and any alternative spellings based on his review of several sources, the earliest date of use, and any explanation evident for identifying each place. These narrative explanations usually fell into natural descriptors, names based on land-ownership, or historic events. The list of resources used to assemble this study begins with the earliest maps available for the region, interviews and "consultants" representing many of the longest-living residents of Dane County's towns, secondary sources that include the earliest published memoirs of the county, published histories by a succession of Wisconsin's most renowned historians, and the comprehensive use of other various sources, including Post Office lists, railroad maps and schedules, and personal confirmation of local uses by current residents.

Cassidy began this study shortly after his arrival at the University of Wisconsin in 1939, after receiving his Ph.D. from the University of Michigan in 1938. He arrived to teach in a department that would become world-renowned for his later scholarship as editor of the *Dictionary of American Regional English.* Encouraged by then Graduate School Dean E. B. Fred (later university president), Cassidy began his study in addition to his teaching duties. Employing graduate students in the task, this research built somewhat on the process that produced a series of sociological studies published as University of Wisconsin Agricultural Experimental Station Research Bulletins. One of the earliest and most influential is the study of ethnic neighborhoods of Dane County conducted by J. H. Kolb in "Rural Primary Groups: A Study of Agricultural Neighborhoods" (Research Bulletin, no. 51 [1921]: 12–13). Kolb identified geographical areas in Dane County based on land titles, names of property owners, and the boundaries of historic ethnic neighborhoods as well as the borders of "contemporary" neighborhoods that existed in 1918. This study was part of the groundbreaking examination of rural sociology that Professor Kolb developed at the University of Wisconsin. In 1922 Kolb and his students began studies of Dane County "drainage service basin" communities and "cultural watersheds" the rural neighborhoods centered on and serviced by mercantile crossroads communities, parish churches of different denominations, and the regional foci of agricultural business. Based on the 1920 census data, Kolb's studies fleshed out attributes of the regional communities described in the census data at a time when agriculture and transportation were undergoing a transition from horse-power to a gasoline-based economy. Kolb expanded his studies with publications on "Service Relations of Town and Country" (1923) and "Special Interest Groups in Rural Society" (1927). Soon thereafter, another University of Wisconsin sociologist, Charles Galpin, published his study "Dane County, Wisconsin" in the United States Department of Commerce, Bureau of the Census, *Farm Population of Selected Counties* (Washington, D.C.: GPO, 1924). By 1929 nearly all Dane County farms had a car or truck, and a decade after the Good Road Movement began its legislative initiatives to improve the delivery of farm goods to market the conversion of county highways from gravel to macadam or concrete was underway. Through these years of transition and into the 1940s, Galpin continued to publish his studies of rural Dane County.

Coincident with Cassidy's arrival, the great urban planning academician Ladislas Segoe was in Madison to create a study to guide the future zoning and urban plan for the city. Studying the city and making recommendations for redefining the zoning laws and establishing a series of developmental criteria for directing certain types of growth to particular regions in the city, Segoe published his study in 1939, before he went on to become an influential author and professor of regional and urban planning.

In many regards, Cassidy's study could not have begun at a better time, especially considering that the literal reshaping of the Madison landscape was in the making from the point of his arrival to the publication of his study in 1947. Despite these landmark studies and the contemporary climate, Cassidy confined his study to the State Historical Society of Wisconsin map and history collections, which held maps from nineteenth-century development in Wisconsin up to the most recent plat maps for geographical place-name origins. Cassidy's list of references, most included in the Historical Society's collection, was made up of an exhaustive list of the most scholarly histories of Dane County and Wisconsin, as well as individual collections by notable residents and interviews with many of the most prominent men and women living in Dane County's towns and villages. Using a scholarly approach that thoroughly encompassed the best reference sources about the county, Cassidy published his work in *Publication of the American Dialect Society* (Number 7, April 1947).

Comprehensive dictionaries of place-names provide curious readers with well-researched information about the place where they live. As fuel costs rise and the Dane County economy continues to undergo its transition to serve a regional population beyond its county borders, Cassidy's study of place-names remains a valuable reference to explain the name origins for Dane County geographical places. Dane County's population will continue to change over time, and its communities will bear less and less resemblance to those extant when Cassidy's study was published. To that end, this work offers readers interested in local history a valuable understanding of the Dane County landscape that existed at the end of World War II. More than forty years after its initial publication by the University of Wisconsin Press, Cassidy's text remains a valuable historical resource. It was useful to me in writing *Forward! A History of Dane, the Capital County*, as much for its list of place-name origins as for the resource list Cassidy culled to identify those names. In his appendixes, Cassidy charts the influence of certain names on other, often adjacent locations and shows how a list of similar elements aids in that naming process. While he selects only a handful of such locations, his analysis provides insight into the nomenclatural process involved in naming places, an added topological asset for others who might undertake this type of study. Additionally, Cassidy's discussion of the topographical topoi realtors use to create names for subdivisions is as entertaining as the ironic names of the subdivisions themselves.

In the New World, place-name studies allotted a cultural value to geographic areas and loci based on their identifying descriptors. Name study presents a resource for local history buffs and scholars alike to understand the influences involved in the names of places in their communities. Cassidy created a valuable resource for understanding the Dane County landscape at the end of World War II, in that most physical geographical and geological features had retained their names. Since then, the greatest change has been the explosion of residential subdivisions named by real estate developers and platted into Dane County's expanding city and village borders. Whether or not Cassidy would have valued these names by including them in his dictionary is a question left to ponder.

INTRODUCTION TO THE 1968 EDITION

GENERAL REMARKS

The following study has been made "in depth"; that is to say, it has sought to collect all the place-names that have or have had any sort of public status (however restricted) within Dane County, Wisconsin, since it was first permanently settled by the white man somewhat over a century ago, and to analyze this collection for the light it may throw upon place-naming in this region.

Though all the major documentary sources and many minor ones have been thoroly studied, and though over 120 informants have been consulted (see the lists in Appendix II), completeness has not been achieved. Nevertheless, it has certainly been approached closely enough so that the basis of evidence may be considered sound.

Great care has been exercised to get accurate evidence, and to indicate how far any evidence used may be trusted. Even the most dependable sources sometimes go astray; for example, the U. S. Geological Topographic maps were caught in error (see *Elvers Cr.*, *Black Bridge Crossing*), as also the State Highway maps (see *Little Sugar River*—corrected recently, *Montjoy*). The writer does not fancy himself exempt from this company, and will be glad to be informed of any corrections that the reader can make.

The major part of the study, then, is the list or "dictionary" of Dane County names. Arranged alphabetically, with full cross-references, it should be easy to use. Obsolete names—over one fourth of the total—are preceded with an asterisk. Not every possible analysis of the evidence has been made; the emphasis has been put upon lexicographical matters. The arrangement of the material, however, will easily permit the reader to make additional analyses—historical or cultural, for example.

THE KINDS OF INFORMATION SOUGHT

The methods of the New English Dictionary (Oxford Dictionary)* have been taken as the model wherever applicable to a place-name study; elsewhere they have been adapted. The result will best be seen thru the form given to the entries. This is also partly modeled on the excellent place-name studies done in Missouri under the direction of Professor Robert L. Ramsay. Following are the kinds of information given whenever possible about each name, and beside them, the more or less corresponding kinds of information given for words in the New English Dictionary. Of course, the pertinent facts about words and names are not exactly the same; besides, the NED covers about twelve centuries, and this list just over one.

Dane County Place-Names	*NED Words*
Present spelling	Head-word (or Lemma)
Type-of-feature label	Part-of-speech label
All variant spellings, with dates	All variant spellings, with dates
Pronunciation of current words	Pronunciation of current words
Type of name	
Etymology or provenience	Etymology
Circumstances and people involved	
References to sources of evidence	Dated quotations from sources
Date-range of the use of the name	" " " "
Degree and level of usage	Usage labels and discussions
Location, spatial range, and changes therein	Meanings and their development
Other names which this feature bears or has borne	Cross-references to related words

Spellings. All those found are given, except where very numerous, in which case typical examples are given and a source for the rest is referred to. Whenever spelling-variants can be dated—which is not often—they are.

Type of feature. This is indicated immediately after the head-word, unless the name already contains it. Thus *Deer Creek*, but

* Henceforth abbreviated NED.

Deer Creek, cor.; *Luhman's Corners*, *Darwin Station*, but *Sugar River Station*, vill.

Pronunciation. This can seldom be given for obsolete names, and is not necessary for *Big Ditch*, *Mud Lake*, and their ilk; but elsewhere pronunciations are included. What appears commonplace in this region may be unusual or unknown in another; so it is safer to err on the side of inclusion. Pronunciations heard from a particular individual are so indicated. The International Phonetic Alphabet is used in a broad transcription, except in foreign-language names, with which a narrower transcription became necessary. (See *Publ. of the Amer. Dialect Society*, No. 1.)

Type of name. This usually follows the pronunciation. Full lists of these types (descriptive, inspirational, etc.) are given and discussed in Appendix I.

Etymology. This has been given only for those names directly from some non-English language. For English-language names etymology is either unnecessary (*Bald Hill*, *Goose Lake*) or does not properly belong to the local name. For example, *Cambridge*, Wisconsin, does not mean the town where there is a bridge over the River Cam, formerly the Granta River; it is on Koshkonong Creek, and the names of the creek and village are quite unrelated. For such a name, then, the question of etymology gives way to that of provenience: this village was named for Cambridge, N. Y. Ultimately this may no doubt be traced to the original Cambridge, England, which name has an etymology. In this study, name origins have usually been traced back only one step.

Circumstances and people involved. These cannot be known in the majority of instances; the information simply has not been preserved. Particularly with natural features, the namer is as nameless as the author of a popular ballad. Sometimes (as with such vapid school names as *Maple Lawn School*, *Oak Hill School*) the authorship might perhaps have been discovered, but did not seem worth the labor. In a great many names, however, formerly accepted explanations have been rejected, doubts clarified, and new solutions offered. (See particularly *Cross Plains*, *Lake Koshkonong*, *Deerfield*, *Pheasant Branch Creek*, *The Four Lakes*, *Token Creek*, *Badfish Creek*.)

Date-range of the use of the name. This presents problems all its own, yet it seemed essential that every bit of pertinent evidence on it should be collected. Without at least approximate dates, such matters as changes in terminology, in styles of naming, in the spreading of names, cannot be studied accurately. In the pages below are several instances of words used in Dane Co. place-names at dates earlier than those recorded in the historical dictionaries. Dated place-names may also be used for studies of settlement history.

Precise dates of naming are known for only a few natural features: *Nine Mound Prairie*, *Badger Mill Creek*, etc. For most others an approximate date can be arrived at from maps, early accounts, the time of the known coming of settlers, etc.: *Island Lake*, *Saunders Creek*.

With man-made features, however, dating is more often possible; with some it may be quite exact—PO's for example. A PO name had to be chosen and approved before the office could be officially established or put into operation; but until the latter stage was reached, the name was not actually a PO name. In cases where PO's were discontinued and re-established many times, the name, as a PO name, has actually lapsed and been revived as many times: *Pine Bluff PO*, *Lake View PO*. When a name remained after discontinuation of a PO, it must be considered as applying to the place at which the PO had been kept: *Acorn*, *Peculiar*.

Other names that depend on official action can also be dated closely: county, town, plat names, etc. Occasionally, a name has pre-existed official action; when this is known, it is recorded. *Baskerville Harbor*, for example, was named several years before the plat was recorded; and the date of record of plats has been taken as official.

Beginning dates are comparatively easy to know about; when a name may be considered obsolete is another matter. Names do not die thru official action. A plat may be legally vacated, or may become absorbed into a city, but the name will survive the change in status of its referent: *Fair Oaks*, *Nakoma*. When a village is moved to a new location, the former place may prefix the word "old" and so keep the name: *Old Deerfield*. Some names are extraordinarily tenacious. In 1855 a law was passed making the present names of the *Four Lakes* and the *Yahara River* official, yet over ninety years later none of the former names is out of use, and the last, *Catfish Creek*, is more popular locally than the official name.

The names which have become obsolete are those which never gained much currency to begin with, which existed on paper mainly (e.g., on early maps), which were extremely local, or were attached to some institution which disappeared. But once a name gained more than local currency, and particularly when some anecdote (not necessarily a true one) became connected with it, it has had far more vitality: *Halunkenburg*, *Whiskey Creek*. In a few instances the name has been kept alive by historically conscious users, even when it is unofficial: *Nauneesha River*.

It is often impossible to know when a name is really out of use, however. As long as it is still known to the oldest generation it is still alive, and may be passed down as a relic, if no more. A feature may thus have several names at once, more or less active: a recent name is *Remy Spring*, but *McFadden Spring(s)* is still well remembered for it, and a few still know of the name *Grand Springs*, tho it is wholly out of use. Only the third is marked obsolete in the list.

References to sources of evidence. These are made by means of abbreviations (listed in Appendix II with the documentary sources, maps, or informants they stand for). They are given at any point in an entry, following the information or statement to which they apply. Not all sources are mentioned; many are quite obvious—county histories, postal guides, etc. References to specific sources are therefore made only when:

a. The source is directly quoted;
b. There is but the one source;
c. Sources are at variance;
d. The source gives new evidence;
e. The source gives further details or discussion of the problem.

Degree and level of usage. As with dates, to know these is also essential to a historical study. How current has the name been? Is it restricted to some group, or is it generally known or used? Has it changed? In line with general dictionary practice, general usage is not mentioned; but when there is a variation of some kind, that is recorded. It may be assumed that the names of most "corners" and "hills" are known only locally; it may not be assumed that the presence of a name on a map proves its local usage: see *Deer Creek*, *Picture Rocks*, etc. Records of whatever kind sometimes give a factitious existence to names, yet since records must be depended upon to some extent, and any record is more than no record at all, even factitious and nonce names have been listed. Many of these are found in early maps, and every available map up to 1850 has been examined. Many statements about usage also rest on immediate observation in the field, and on the word of informants.

Location, spatial range, and changes therein. For a historical study, these facts are also essential, since the application of place-names is not stable. To be strictly logical, every change in the referent makes a change in the meaning of the name. But to record every such variation—for example, the increases in the corporate area of the city of Madison—would hardly be valuable, and has not been attempted. On the other hand, if the semantic value of the generic parts is to be known, just what they are applied to must also be known. What, for instance, is the local distinction between "lake" and "pond," "river" and "creek"? To answer such a question, one must know the physical facts about the features to which each name is applied.

The peripatetic nature of early PO's is discussed below. Schools were also moved about. Both the location and area to which some town names were applied changed in the most puzzling fashion—e.g., the area now comprised by the towns of Berry, Roxbury, Mazomanie, and Black Earth. To parts of it, at various times, have been applied the names of *Mazomanie* and *Ray* (area changed once), *Dane* and *Black Earth* (area changed twice), *Roxbury* (area changed three times), and *Farmersville* (area changed four times). To know what "town" has meant in this area one must know these facts, among others.

Early maps and many more or less authoritative accounts frequently misapply place-names, particularly those of streams: *Pine River*, *Duck Creek*. The part of a valley, river system, etc., to which a name is applied may change in time: *Catfish Creek*, *Halfway Prairie*. An island may become a peninsula: *Governor's Island*. A single stream may simultaneously have several names applying locally to different parts, and thus overlapping with the name or names of the whole: *Yahara River*, *Catfish Creek*, *Whiskey Creek*, *Holum's Creek*. In all these cases the referent of the name has been mistaken, has changed, or has in some way fluctuated; correct locations and changes in application must be found out and recorded.

Other names which this feature bears or has borne. These are simply listed as cross references, usually at the end of an entry.

ABBREVIATIONS

C&NW—Chicago & Northwestern Railroad
CMSP—Chicago, Milwaukee, St. Paul & Pacific Railroad
cor.—corner
DAE—Dictionary of American English (Univ. of Chicago Press)
disct.—discontinued
est.—established
IC—Illinois Central Railroad
NED—New English Dictionary (Oxford Univ. Press)
quad.—quadrangle
R—range (surveyor's sense); see below
re-est.—re-established
sec.—section (surveyor's unit)
subd.—subdivision
T—township (surveyor's term); see below
town—township (legislative sense)
vill.—village
WHS—Wisconsin State Historical Society
*—(preceding a name) out of use
T7N,R12E (and other such combinations)—the conventional surveyor's abbreviation for *township 7 north*, in *range 12 east* of the fourth principal meridian.
Quarter-sections are not mentioned, but are implied by the use of compass directions with section numbers: SE sec. 4 would mean the *southeast quarter of section 4*.
Capital letters are used for town names: Madison, the city; but MADISON, the town.
All cross references are in italics.
Abbreviations of sources are listed with them, in Appendix II.

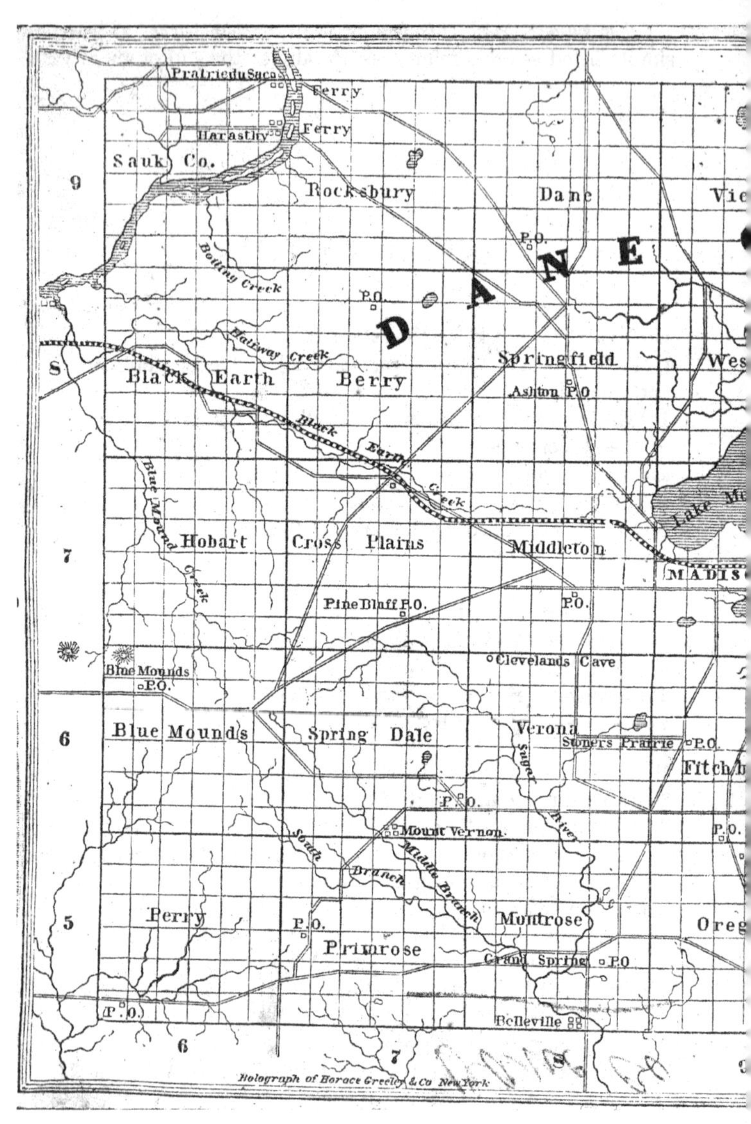

Dane County, 1855 (Wisconsin Historical Society, WHi-65043)

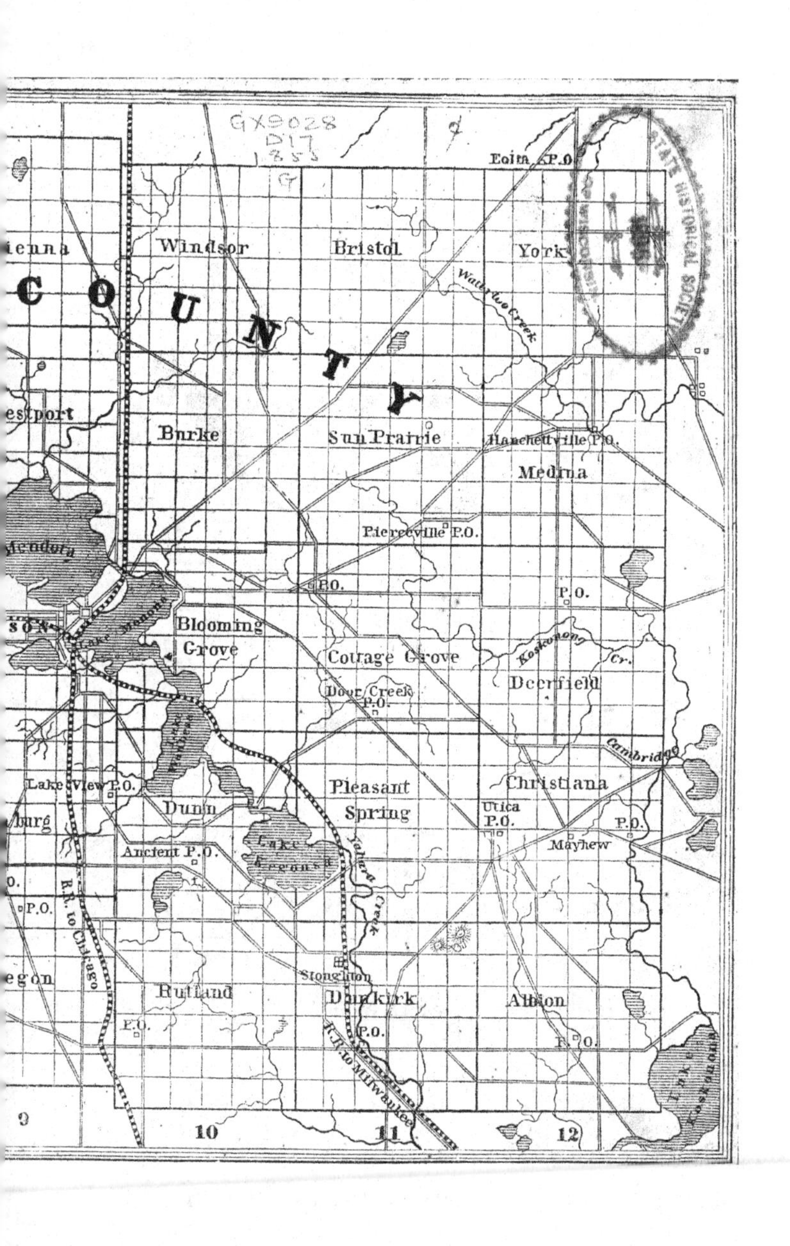

ienna
Windsor
Bristol
York
COUNTY
Eolia
Waterloo Creek
estport
Burke
Sun Prairie
Hanchettville P.O.
Medina
Mendota
Pierceville P.O.
P.O.
Blooming
Grove
Lake Monona
Cottage Grove
Koskonong Cr.
Deerfield
Door Creek P.O.
Lake Waubesa
Cambridge
Lake View P.O.
Pleasant
Spring
Christiana
Dunn
Utica P.O.
burg
Mayhew
Ancient P.O.
Lake Kegonsa
Yahara Creek
R.R. to Chicago
egon
Stoughton
Rutland
Dunkirk
Albion
R.R. to Milwaukee
Lake Koskonong
9
10
11
12

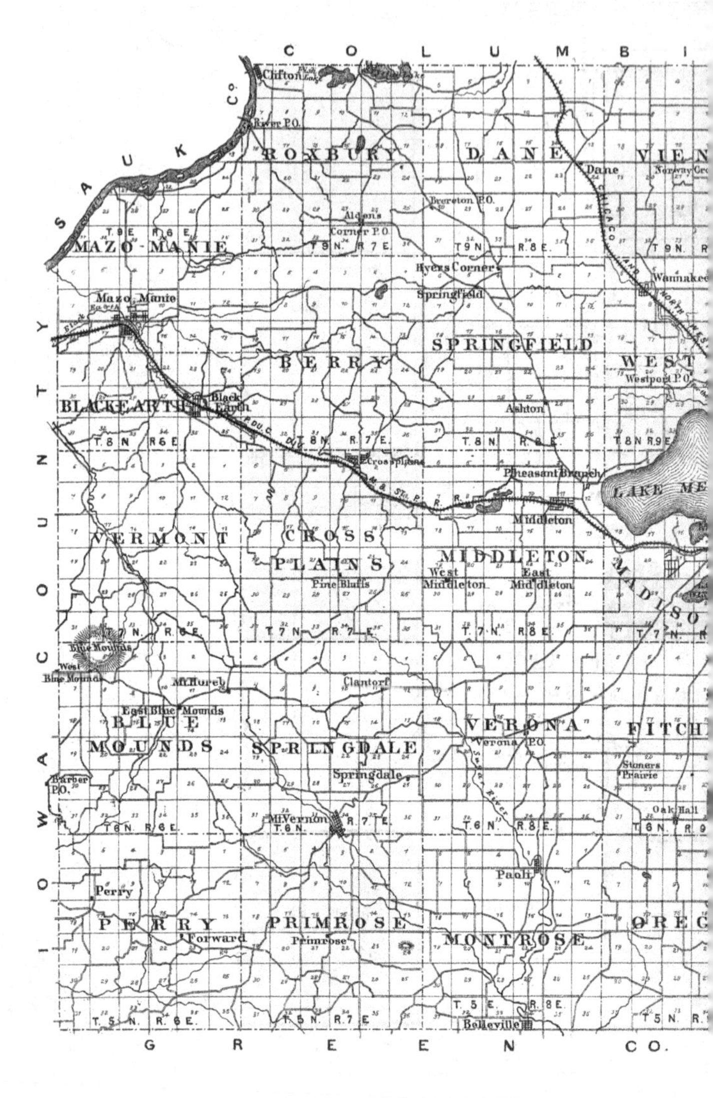

Dane County, 1873 (Wisconsin Historical Society, WHi-65474)

e Map of

CO. WIS.

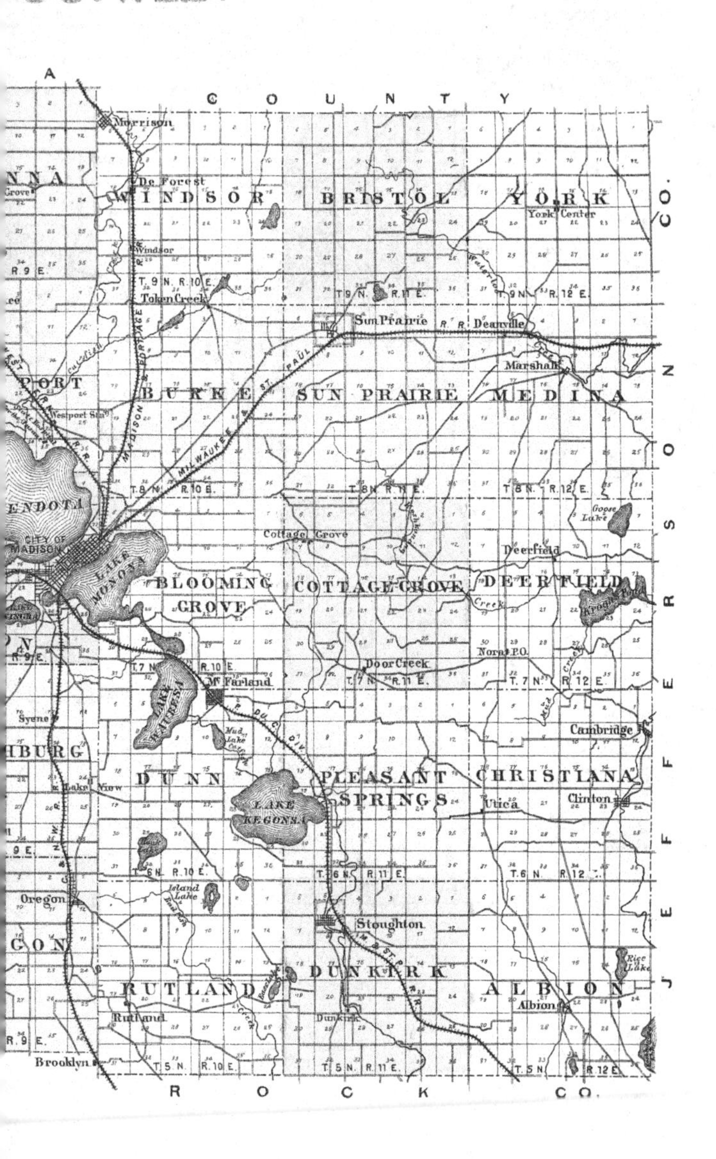

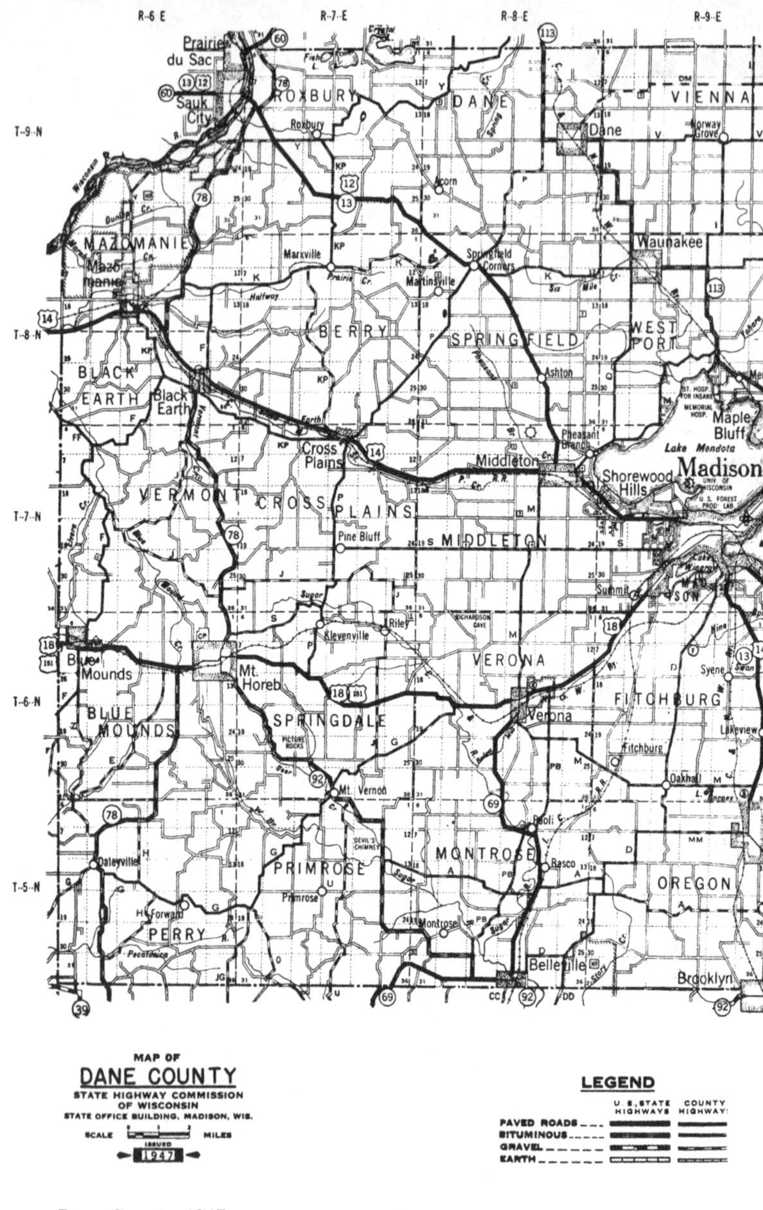

Dane County, 1947

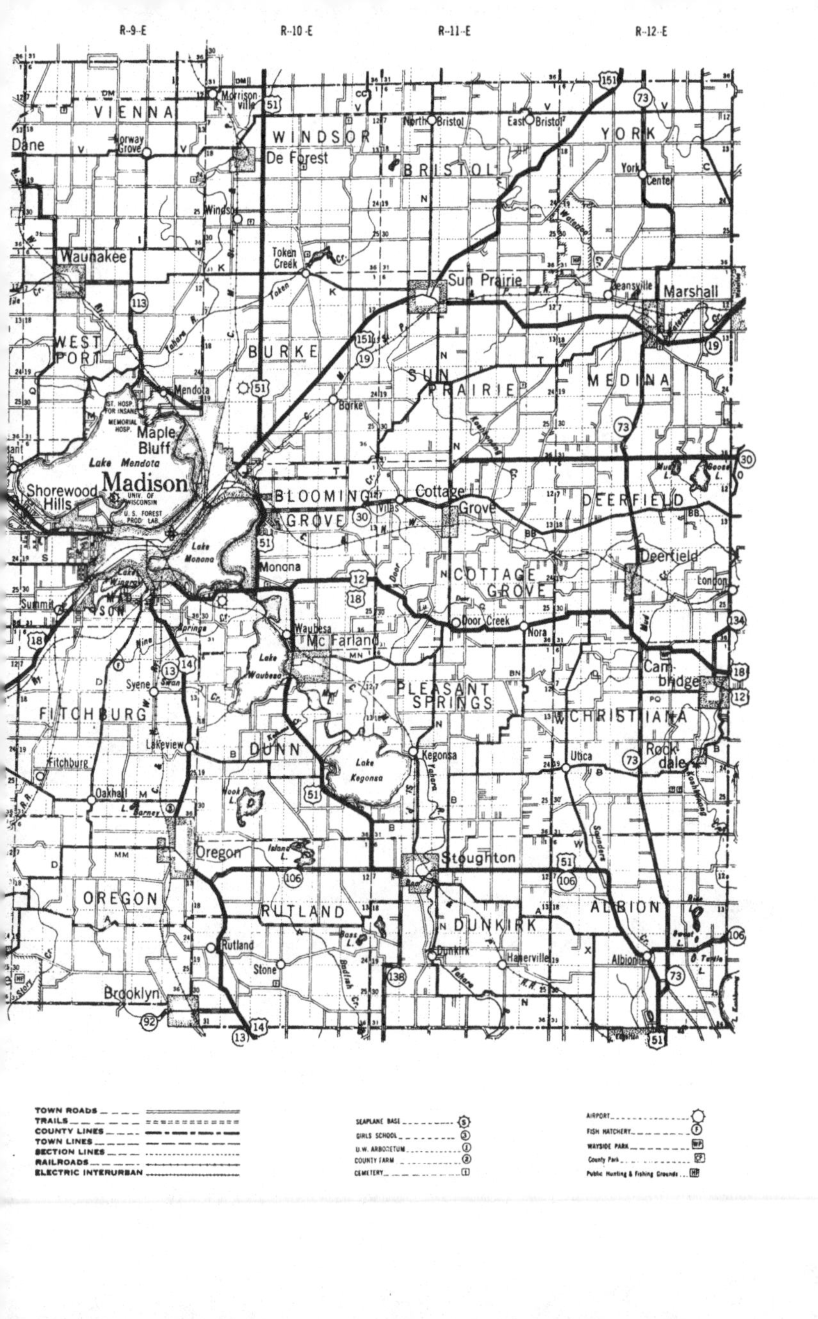
R-9-E
R-10-E
R-11-E
R-12-E
VIENNA
WINDSOR
BRISTOL
YORK
Dane
Norway Grove
Morrisonville
De Forest
North Bristol
East Bristol
York Center
Windsor
Waunakee
Token Creek
Sun Prairie
Deansville
Marshall
WEST PORT
BURKE
SUN PRAIRIE
MEDINA
Mendota
Burke
Maple Bluff
Lake Mendota
Madison
Shorewood Hills
UNIV. OF WISCONSIN
U. S. FOREST PROD. LAB.
ST. HOSP. FOR INSANE
MEMORIAL HOSP.
BLOOMING GROVE
Cottage Grove
Vilas
DEERFIELD
Lake Monona
Monona
Lake Wingra
Deerfield
London
COTTAGE GROVE
Summit
Waubesa
Mc Farland
Door Creek
Nora
Lake Waubesa
Syene
FITCHBURG
PLEASANT SPRINGS
Cambridge
CHRISTIANA
Lakeview
DUNN
Kegonsa
Lake Kegonsa
Utica
Rockdale
Fitchburg
Oakhall
Oregon
Stoughton
OREGON
RUTLAND
DUNKIRK
ALBION
Rutland
Stone
Dunkirk
Hanerville
Albion
Brooklyn
TOWN ROADS
TRAILS
COUNTY LINES
TOWN LINES
SECTION LINES
RAILROADS
ELECTRIC INTERURBAN
SEAPLANE BASE
GIRLS SCHOOL
U.W. ARBORETUM
COUNTY FARM
CEMETERY
AIRPORT
FISH HATCHERY
WAYSIDE PARK
County Park
Public Hunting & Fishing Grounds

Dane County Place-Names

Acorn, cor. ['ekən]

For the PO, and at the same location. No longer widely known.

*** Acorn PO**

Named by Mrs. Kate Goddard, the postmistress, tho why is unknown; she seems to have submitted to the PO dept. a list of alternatives, from which this was chosen. If there was any hope that from this little acorn a great oak would grow, this hope was disappointed, since the PO lasted only from Sept. 17, 1888 to July 13, 1898, and no village grew up. It was kept at the postmistress' house, NW sec. 30, DANE.

Acorn School

For the nearby *Acorn PO* and corner: at the road junction in SE sec. 30, DANE. Closed for many years.

*** Adam Smith Tavern, The**

For its owner: a tavern (the "Burke Cottage Hotel") in NE sec. 14, BURKE, operated from 1843 to 1897 by Adam Smith of N. Y., prominent early settler. (Butt. 1288)

*** Adsit PO** ['ædsɪt]

For Stephen H. Adsit and his family, from N. Y. state, who settled (1846) in sec. 5, DEERFIELD. Here the PO was kept when it was est. June 8, 1882; it was moved before 1886 to SW sec. 8; disct. Oct. 31, 1899.

Albion, town ['ælbɪən]

For Albion, Orleans Co., N. Y., the former residence of Isaac Brown, an early settler, who suggested the name. He was supported by some English settlers to whom it also suggested home. The town, when first formed Feb. 2, 1846, included the present CHRISTIANA; when this was separated Jan. 3, 1847, Albion was reduced to its present area; T5N, R12E.

Albion, vill.

For the town. First settlement within its area was in 1842, but it was not platted till 1944. It includes adjoining parts of secs. 21, 22, 27, and 28, ALBION. Formerly also *Albion Center.*

*** Albion Center,** vill.

For its position near the center of the town: an early alternate name for *Albion* village.

Albion Creek

For the town and village of Albion, thru which it flows: an alternate local unofficial name for *Saunders Cr.*

Albion Marsh

For the town and perhaps the prairie: a large marshy region, chiefly E of Saunders Cr., in the center of ALBION, but also applied to the marsh W of the village. (Clough)

Albion PO

For the town. Est. Feb. 11, 1847, and kept on sec. 22, ALBION, where the village later came to be.

Albion Prairie

For the town: actually this is a southward spur of *Koshkonong Prairie;* it is largely in the NE part of ALBION, between Koshkonong Cr. and Saunders Cr., and runs up thru the S part of CHRISTIANA (which was once part of ALBION). Early settlement in the town began on this prairie.

Albion Prairie School

For its location on Albion Prairie, NW sec. 14, ALBION.

* **Albion Precinct**

Park (545) mentions an election precinct by this name as having been established between 1839 and 1846. Since none such is mentioned in Dane, however, Park was probably referring to the *Clinton Precinct.*

* **Alden's Corners** ['ɔldənz]

For John and B. Alden, who owned the land on all sides of it: the crossroad where secs. 26, 27, 34 and 35 meet, ROXBURY.

* **Alden's Corner School**

For *Alden's Corners,* at which it was. Now closed.

* **Alden's Corners PO**

For the corners. Est. May 4, 1860; disct. May 7, 1879.

Allis Heights, subd. ['ælɪs]

For Frank W. Allis, owner of the land from the 1890's; platted 1927, in NE sec. 16, BLOOMING GROVE.

* **Allis Pier,** landing

For Frank W. Allis, owner: one of the stops on the rural mail route in L. Monona, shown on Rural map (1910); SE sec. 8, BLOOMING GROVE.

* **Ames School** [emz]

For John N. Ames, of N. Y., and his family, who settled in N sec. 22, in 1845, and on whose land the school was built: at the road junction between secs. 15 and 22, OREGON. Also known as *Chase School* and *Babbitt School.*

*** Amos PO** ['eˌmɑs]

For Amos Babcock, a friend of the first postmaster, Henry Humphrey, in whose store it was kept, at the crossroads between secs. 34, CHRISTIANA, and 3, ALBION (Bab.). Est. May 20 1893; disct. Dec. 31, 1903. (Gay, 1899, shows it in sec. 27, CHRISTIANA, but this is probably an error.)

*** Ancient PO**

Origin unknown. Name changed from *Dunn PO*, Jan. 14, 1850; disct. May 19, 1850. Kept by Calvin Farnsworth, on his land, N sec. 28, DUNN.

Andrew Henry School

For Andrew Henry, an early settler here (before 1861). The school is on his former land at the crossroads in sec. 23, VERONA.

Andrews Corners ['ænˌdruz]

For Justus Andrews, of N. Y., and his descendants, who have owned adjacent land since before 1873: the crossroads between secs. 24 and 25, YORK.

*** Angleworm Station,** landing (also, **Angle Worm** ——) ['æŋgl̩ ˌwɝm]

A nickname applied by Capt. Frank Barnes to his steamboat landing on L. Monona, at the foot of S. Carroll St., Madison, at which he delivered a humorous lecture on angleworms every fourth of July for many years. He operated his boats from about 1854 to 1902, but the name is still found on a map in 1910 (Rural).

*** Anthony Creek** ['ænθənɪ]

For David Anthony, on whose land (NE sec. 19, RUTLAND) the stream rose. He settled in 1840; the name was used as late as the 1880's. The stream flows E and a bit N to join *Badfish Cr.* on sec. 16. Also called *Rutland Branch.*

Anthony Springs

As the source of *Anthony Creek:* NE sec. 19 and NW sec. 20, RUTLAND. Called "Anthony's Spring" in 1845 (Dane 126).

Arboretum, The University of Wisconsin

The word "arboretum" has its usual sense here: wooded lands belonging to the University of Wisconsin. Est. 1932, they now include parts of secs. 27, 28, 33, and 34, MADISON, and 4, FITCHBURG.

Arlington Heights, subd.

Reminiscent, and also chosen for its suggestion of elegance or distinction. Platted 1927 within the city of Madison.

Ashton, cor.

For *Ashton PO,* which was moved here in 1867: the crossroad in W sec. 26, SPRINGFIELD. Also called *Ashton Corners.* Formerly *Pickarts' Corners.*

*** Ashton Church,** cor.

For its having *St. Peter's Church* here, and being near the former *Ashton PO.* Shown thus on USGS 1906–7. Now *St. Peter's.*

Ashton Corners

= *Ashton.*

*** Ashton PO**

Probably for Thos. Ashton, President of the British Temperance Emigration Society (see *Gorstville*), some of whose members settled here, among them George Gillett, 1846, in S sec. 22, SPRINGFIELD. Here the PO was est. Sept. 7, 1849, the postmaster being Henry Gillett; it was next moved to Smith's tavern, at the junction in N sec. 22, until 1867, when it was again moved to *Pickarts' Corners.* It was disct. Oct. 31, 1903.

Ashton School

For nearby *Ashton* corner, the school being three quarters of a mile W.

*** Assembly PO**

For the *Monona Assembly,* which it served. Est. July 15, 1897; name changed to *Fair Grounds* Oct. 13, 1899.

*** Aztalan Highway, The** ['æztəˌlæn]

For Aztalan, Jefferson Co., thru which it passed. It was surveyed 1840, and built shortly after; the name lasted in use at least into the 1870's (Eldred). In Dane Co. its course was from the county line at Waterloo, thru Marshall to Sun Prairie—now State highway 19. It was also called the *Sun Prairie-Lake Mills Rd.*

*** Babbitt School**

For Rosel (or Roswell) Babbitt and family, who in 1845 opened a farm adjoining the crossroads at which this school later was. Also known as *Ames School* and *Chase School.*

Back Bay, subd.

For its position behind (i.e., W of) *West Bay* and *Monona Bay;* probably also reminiscent, for example, of the high-class suburb of Boston; platted 1902; since absorbed into the city of Madison.

Badfish Creek (also, **Bad Fish,** till at least 1836; formerly, —— **River**)

Traditionally said to have been the exclamation of an Indian who thought he was poisoned by eating some fish caught here. The frequently printed version of the story (Butt. 863–4, etc.) is suspiciously romantic, with Indian lovers whose tribes are enemies, but who have a secret rendezvous. The young man eats some fish and becomes sick. This part may be the factual basis, if there is any; but the whole story is probably an invention.

It is much more likely that Badfish arose as an alternate of Catfish: the Ojibwa *mān-ameg,* meaning "catfish," if translated quite literally, would be "bad (i.e., small, petty, less desirable) fish" (Geary). Since Badfish Cr. is a tributary of Catfish Cr. (the Yahara R.), early mapmakers may in this instance (as in many others) have considered the two as parts of the same stream; variant translations of the same Indian word might then have been put upon the two branches, and the identity of source have remained unknown or been forgotten.

It is noteworthy therefore that the name "Badfish Cr." antedates the government survey of 1833, being first used on Farmer's map of 1830; here it is in two words, hinting at the literal translation, "Bad Fish." "Catfish" first appears on the surveyor's map (1833), on which Badfish Cr. is simply labeled "a deep creek." All of these facts support the explanation just offered.

The single-word form of the name appears first on Lapham's map (1855).

The present *Oregon Br.* was at first indicated by mapmakers as the real source (it is the longer one); later, *Anthony Cr.* (*Rutland Br.*) was so indicated; maps differ, and today the tendency is to leave the branches unnamed. They join at *The Forks,* whence the creek flows SE into Rock Co.

Badger Mill Creek ['bædʒɚ]

For the Badger Mill, built on this creek in 1844—the first grist mill in the county. ("Badger" here surely refers to Wisconsin and its people.) The name first appears in print in the H & W platbook, 1873. The creek rises in sec. 13, VERONA, and flows SW to join Sugar R. on sec. 28. (Durrie abbreviates it to Badger Creek, 410.)

Badger Mill Hill

For the Badger Mill, built there 1844: a slope over the N bluff of Badger Mill Cr., N sec. 22, VERONA. (Davids.)

*** Badger Mills**

For Wisconsin, the "Badger" state: built 1844 by William A. Wheeler and George Vroman; the first gristmill in Dane Co., the village of *Verona* grew around it.

*** Badger Mills Precinct**

For its location at the *Badger Mills* settlement: an election precinct, est. Oct. 24, 1844, and including the present VERONA.

Badger Park, subd.

For "patriotic" reasons, or to gain favor, the badger being the animal associated with the state of Wisconsin, and a Badger being a Wisconsinite, or a person connected with the University of Wisconsin. Formed out of *Quarrytown;* platted 1917; NW sec. 21, MADISON.

*** Badger Prairie**

Probably for *Badger Mill Cr.;* the prairie thru which this creek flowed stretched diagonally thru the middle of VERONA. (Only in Durrie, 410.) See *Mill Creek Prairie.*

Badger School

For the animal symbolic of Wisconsin. At the crossroads between secs. 16 and 21, COTTAGE GROVE. Now closed.

*** Baertown** ['bɛɚˌtaʊn]

= *Baerville.* (Dahmen)

*** Baerville** (also, **Bearville**), vill. ['bɛɚˌvɪl]

For William, Ferdinand, and (chiefly) John Baer, brothers, who settled here about 1851. John ran the store, and made the original plat called *Cross Plains* here; he died 1877, but his descendants continued in the locality. The name has fallen into disuse, however (Dahmen). Known locally as "upper town."

Bailey Neighborhood, The

= *The Bailey Settlement.*

*** Bailey Settlement, The**

For Samuel, Asahel, and Charles Bailey, who, with others, made an early settlement (1841) in secs. 19 and 30, SUN PRAIRIE. Now called the Bailey Neighborhood.

Bailey School

For the *Bailey Settlement,* at the crossroads of which it is. Now closed.

Baker School

For *Baker's Corners*, at which it is. Now closed.

*** Baker's Corners**

For Ephraim Baker, of Mass., and his family, who from 1844 till after 1890 owned adjoining land: the crossroads where secs. 3, 4, 9 and 10 meet, BRISTOL.

Bakke Hill ['bɑkɪ]

For P. O. Bakke and his descendants, who owned the land from before 1890 till after 1911: a hill in SW sec. 17, PLEASANT SPRINGS.

Bald Hill

Because it has a bare spot on the top of its N side: a large, wooded hill in NE sec. 36, VERONA, and NW sec. 31, FITCHBURG. So called for over 65 years (Gannon).

Baldwin Pond

For Anson and Julius Baldwin, who settled the adjoining land (before 1861): a sizable pond in S-central sec. 26, RUTLAND. (Ames)

Balke School ['bɔlkɪ]

For M. Balke and his descendants, who have owned the land on which it is from before 1873. The present official name, however, is *Pleasant Hill School.*

*** Baraboo Air Line Railroad** ['bærəˌbu]

For its terminus in Sauk Co.: a former name of a branch of the C&NW railroad. It runs N from Madison thru Waunakee and Dane, leaving Dane Co. between secs. 2 and 3, DANE. It was built in 1871.

Barber's Bay

For George Barber and family. Barber bought the land in NW sec. 26, DUNN, in the 1890's; see *Barber's Hill* and *Crown Point.* The bay is on the W side of L. Kegonsa, between *Colladay's Point* and *Lund's Point.*

Barber's Hill

For George Barber and family; see *Barber's Bay:* a spur or ridge in NW sec. 26, DUNN; the E end runs into L. Kegonsa, forming *Crown Point.*

*** Barnes' Landing**

= *Angleworm Station.* (Illustrated in Park, 181)

*** Barnes Tavern, The**

For Harry and Capt. Joel Barnes, propriotors: a well known

early tavern at *Middleton Junction* from 1847 into the 1860's. (Barton 1106)

Barwig, cor.

For the PO; the name has continued in platbooks, along with West Middleton, from 1899 forward, tho no longer in actual local use.

*** Barwig PO**

Origin unknown. There is no other place by this name, it seems, and no record of any person of this locality who bore it. It may be an arbitrary coinage, such as were often necessary to avoid duplication of PO names. Est. Apr. 26, 1894, to succeed *West Middleton PO;* the postmaster was Emanuel Showers; disct. Aug. 15, 1900.

Basco, vill. ['bæs,ko]

The current name for what is officially *Basco Station.*

Bascom Hill ['bæskəm]

Because it is crowned by Bascom Hall, named for John Bascom, early president of the University of Wisconsin (1874–1887); a current student name for *College* or *University Hill.*

Basco PO

For Basco, Hancock Co., Ill. When a name was needed for the new PO, and those suggested were found to be already in use, Albert Bavery, a local workman, suggested this. Since contact had been kept up between French residents of the Illinois town and the French settlers in the E part of the town of Montrose, this suggestion had some point, and when the postmistress found that there was no other Basco in Wisconsin, it was accepted by vote (Henry). It was est. Sept. 18, 1889, and kept at *Paoli Station.*

Basco Station, vill. (also, simply, **Basco**)

For Basco PO. When the IC railroad came thru MONTROSE (1888), the station established on sec. 14 was platted and called *Paoli Station.* The next year, a PO was est. here by the name of Basco. Thus Basco PO was kept at Paoli Station, with some resulting confusion, until the latter was renamed Basco Station (1895). The station has been discontinued, tho it is still listed in railway guides.

Baskerville Harbor, subd.

For adjoining Baskerville Park, and E. J. Baskerville, former owner of the land: the entry of Pheasant Branch Cr. to L. Mendota, platted 1941; SE sec. 1, MIDDLETON. The name was in use for more than 10 years before the platting, however.

Baskerville Park, subd.

For E. J. Baskerville, owner of the land; on L. Mendota, platted 1909, in SW sec. 6, MADISON.

Bass Lake [bæs]

Because in early days it contained many black bass: a small lake in NE sec. 24, RUTLAND. Shown without name on maps from 1835 (Terr.) forward; first with name 1849 (Lapham).

*** Bass Lake PO**

For nearby *Bass L.* It was moved when the name was changed from *Stoughton PO*, Apr. 9, 1850, and was kept at the home of the postmaster, Henry Edmunds, in SW sec. 12, RUTLAND; disct. Aug. 15, 1851.

Bass Lake School

For the nearby *Bass L.* In NW sec. 19, DUNKIRK.

*** Battle Ground**

As the site of the Battle of *Wisconsin Heights*, in the Black Hawk War, 1832. So marked on maps from 1835 (Terr.) thru 1850. The battle was fought on parts of the present secs. 19, ROXBURY, and 24, MAZOMANIE; the maps show the latter.

Baywood, subd.

Descriptive, for woods along a small bay in the E shore of L. Mendota; platted 1922; since absorbed into *Maple Bluff*.

*** Beanville Station**

For the many beans grown on the farm of Joseph Gillett, across from it: a loading station on the IC railroad in NW sec. 17, FITCHBURG. The namers, a group of local farmers, having chosen this humorous name, immediately erected a signboard displaying it prominently; thus it became a local byword. It was in existence from about 1910 to 1935 (Vroman).

*** Bearville**

= *Baerville*.

Beaty Hill ['betɪ]

For George Beaty, of Penna., and his family; he settled here 1854: a hill in sec. 36, VERMONT, over which the present state highway 78 goes.

Beaty School

For the Beaty family (see *Beaty Hill*): on the extreme N edge of sec. 36, VERMONT. Now closed.

*** Beaumont,** vill. ['bo,mɑnt]

From French, meaning "beautiful mount," in reference to the nearby *Blue Mounds*, but probably also reminiscent: a village, platted Oct. 29, 1836, by John C. Kellogg and George W. Mower (or Moore), in the S half of SE sec. 6, BLUE MOUNDS, within the limits of the present Blue Mounds village.

*** Beaumont Hollow**

For *Beaumont*, the early settlement which included it: a ravine on the extreme S edge of SE sec. 6, BLUE MOUNDS, running up behind the schoolhouse. So called by children 75 years ago. (MacK.)

*** Beaumont PO**

For the village. Name changed from "Moundville PO" May 18, 1839; changed to "Blue Mound PO," Feb. 9, 1843. See *Blue Mounds PO*.

*** Beecher's Place** ['bitʃɚz]

For Amos Beecher, owner of the cottage in the grove which gave its name to *Cottage Grove*, of which this was an early name. (Park 553)

Belda School ['bɛldə]

For the Belda family, owners of the adjoining land; Joseph Belda settled here before 1873. At the road junction in NW sec. 10, WINDSOR. Also called *Happy Hour School.*

*** Belle Fountaine** (also, **Belle Fontaine, Bellefontaine**)

Anglicized from the French, meaning "beautiful fountain"—a name commonly adopted in the US—and said to have been applied in 1833 by Capt. Brown and Col. Dodge to what was later called *Livesey's Spring* (Keyes 28). This name never gained much currency. Later writers respelled it in French spelling. The early spelling implies an anglicized pronunciation (cp. Bellefontaine, Ohio, which is pronounced [,bɛl'fɑuntən]).

Belle Isle, subd. ['bɛl'ɑɪl]

Ultimately from French, meaning "beautiful island," but also reminiscent; surveyed and named by Ray S. Owen, 1912; platted 1928, in NW sec. 20, BLOOMING GROVE, on *Winnequah Point*, jutting into L. Monona. The outer part of the point has been cut by artificial canals (called *Lagoon du Nord, Lagoon du Sud* and *Sumac Lagoon*) to form two islands, whence the applicability of the name.

Belleville, vill. ['bɛl,vɪl]

For Belleville, Ontario, Canada, the birthplace of John Frederick, earliest settler here (1847). The village was laid out in 1848, platted 1851, incorporated 1892. Originally in S sec. 34, MONTROSE; additions have been made in Green Co.

Belleville PO

For the village: est. Oct. 21, 1852.

Bellevue Park, subd. ['bɛl,vju]

Subjectively descriptive (*belle vue* is French for "fine view"), and reminiscent; on L. Monona, platted 1909, in NW sec. 25, MADISON.

*** Beloit and Madison Railroad**

For the main towns along it: a former name for a branch of the C&NW railroad. It enters Dane Co. on sec. 31, RUTLAND, and runs N thru Oregon. It was completed to Madison in 1864. (Barton 794)

Bergen's Island ['bɝ,gɛnz, 'bɝgənz]

For the owner: a recent alternative name for *Railroad Bridge Is.* (Derl.)

Berg Hill, (The) [bɝg]

For John I. Berg (or Berge), Norwegian settler here in 1846, and his descendants. (Gilb.) It is in the middle of sec. 8, SPRINGDALE, N of highway 18, with a town road coming over it.

Berg School

For the Berg family (see *Berg Hill*). The school is close to the road junction between secs. 5 and 8, SPRINGDALE.

*** Berg Spring**

For James Berg, who owned the land from before 1861 till after 1873: a former name for *Henry Spring.*

*** Berk PO**

For BURKE, tho whether the spelling was changed by accident or by intent is unknown. Est. Mar. 12, 1852, with Abner Cady as postmaster, and presumably kept on his land, SE sec. 16, along the stage route that passed thru Token Creek. Disct. June 15, 1854. See *Burke PO.*

Bermanville, vill. ['bɝmən,vɪl]

For Frank and Gene Berman, owners of the land: a small group of houses in NE sec. 36, FITCHBURG. The name is unofficial and local; the allotment was not made till after 1931.

Bernards' Hill [bə'nardz]

For J. Bernards and his family, owners from before 1874 till after 1904 of the land on which the brow of this prominent hill is: E sec. 36, SPRINGFIELD. Later, *Vasen Hill.*

*** Bernard's Landing**

For William P. Bernard, owner: a boat-landing on the L. Mendota shore, in block 260, city of Madison; in use before and after 1911 (Cant.).

Bernards' Spring

For the Bernards family; see *Bernards' Hill.* The spring issues just below the hill, and flows into *Pheasant Branch Creek.* Later, *Vasen Spring.*

Berry, town ['bɛrɪ]

Probably for Berry Haney, the second settler of the adjoining CROSS PLAINS. Haney was an important early character hereabouts, having opened a tavern (*Haney's* or the *Haney Stand*) as early as 1836. He had the *Cross Plains PO* established, 1838, and was its first postmaster. He opened a ferry across the Wisconsin R. at Sauk City later the next year, returned to Cross Plains 1840, and was in this neighborhood till about 1857 (Dahmen). BERRY (T8N,R7E) was established by separation from *Farmersville*, Jan. 7, 1850.

*** Berry PO**

For the town. Est. May 26, 1851; disct. Nov. 20, 1863. It was kept first by Joseph Bowman (in sec. 7?–Dahmen); before 1861, however, it was moved to about the site of the present *Marxville.*

Betlach's Hill ['bɛtlaks]

For Rudolph Betlach, on part of whose land it is: a small conical hill in N sec. 10, SUN PRAIRIE. Betlach took up this land in the 1890's.

*** Beverly PO**

Origin unknown; if named for some other place, there were at least 5 in existence before this. Est. Feb. 8, 1849, disct. Oct. 30, 1850; re-est. Feb. 10, 1853, disct. Mar. 19, 1857; re-est. June 2, 1858, disct. Dec. 24, 1859. Its exact location is unknown, but it was in BRISTOL, on the Sun Prairie-Columbus Road (present U. S. highway 151).

Big Ditch

Descriptive: a drainage ditch dug about 1910, connecting Dunlap Cr. with the Wisconsin R., running from NE sec. 34 westward thru secs. 33, 32, and 31, MAZOMANIE. It thus carries off some of the water of the creek; Hiway maps even label the ditch with the name of the creek. This name is used only on USGS map, 1916–20.

Big Door Creek

The longer branch of *Door Cr.*, which is the official name of the stream (*Little Door Cr.* being considered a tributary). "Big Door Cr." is found in platbooks, etc., up to 1890; tho not official, it is still in use locally.

*** Big Hill**

Descriptive: a prominent hill in sec. 34, WINDSOR, an early landmark for travelers. So called till at least 1877; now *Token Creek Hill.*

Biglow School ['bɪgˌlo]

For Harvey Biglow, of Vermont, and his family, who settled in S sec. 29, RUTLAND, in 1845. At the crossroads between secs. 29 and 32, RUTLAND. Now closed.

*** Big Marsh, The** (MIDDLETON)

Descriptive: a large peat bog, in secs. 2, 3 (mostly), 10, and 11. So called chiefly in the latter 1800's; formerly *Slaughter's Marsh.* Now drained.

*** Big Marsh, The** (RUTLAND)

Descriptive: an early name for *Island Lake;* used in 1845 (Dane 124).

Big Spring

Descriptive; see *Indian Spring*, Madison.

Big Spring, The

Descriptive. Noteworthy as the site of the first settlement near what became Mt. Vernon; made (1846) by George Patchin and family; E sec. 34, SPRINGDALE.

Big Stone, The

Descriptive: a large piece of gray stone lying isolated on a meadow, in the E part of sec. 28, PERRY. It gave its name to a cheese factory. (Kitls.)

*** Big Woods, The**

Descriptive: an early name for *Vilas Woods* (Brown).

*** Birch River**

From the presence of birch trees? There are many on the Wisconsin River bluffs. Apparently a former name for *Mounds Creek* tho used only on Taylor's map of 1838.

*** Bird's Ruins**

For Zenas Bird, of Little Falls, N. Y., first settler of MEDINA, whose house (built in 1837 on sec. 10) and timbers brought to construct a mill were burned in October 1839 by a prairie fire. The "ruins"—the frame of the house—stood till 1845, giving the name to the locality about them for a good many years. Shown on early maps (J&E, etc., sometimes merely as "Ruins") tho not always accurately placed. The settlement here later became *Hanchettsville*, *Howard City*, and finally *Marshall*.

Bitsedalen, valley ['bɪtsəˌdɑˑlɛn]

Norwegian for "Bates Valley" (Haugen). Said to have been named for an "English" (i.e., English-speaking) family that settled early in SW sec. 14, VERMONT (Gilb.), in what soon became a solidly Norwegian community. No written record of any such family has been found, however. The valley is chiefly in secs. 15, 14, and 11; thru it runs a small tributary of Vermont Creek.

Bittersweet Island

For the bittersweet vines growing there: a small island in the Wisconsin R., E of the lower part of Railroad Bridge Island (Derl.). Since USGS (1892) does not show it, it must have developed out of the swamp since then. In the extreme NE corner of MAZOMANIE.

Black Bridge (Crossing) ['blækˌbrɪdʒ]

Because the wooden bridge built there in 1869 was originally painted black: a crossing of the Yahara R. just below the outlet of L. Monona, in SW sec. 20, BLOOMING GROVE. Town records call the bridge "Catfish Bridge," but in general usage it has been "Black Bridge" from very early. The present bridge is of steel. (Femr.) Originally, an Indian trail crossed here; see *Grand Crossings*. (The USGS map, Madison Quadr. 1904, calls this "Monona Crossing," and puts "Black Bridge Crossing" erroneously below L. Waubesa, in sec. 10, BLOOMING GROVE.)

Black Earth, town ['blæk'ɝθ]

For the creek, PO, and village. The name was first used of T8N,R6E to replace "Farmersville," Feb. 1, 1851. But this area was reduced by loss of its S half when RAY was organized, May 17, 1858. Finally, on Jan. 15, 1859, an exchange was made: "Black Earth" was substituted for "Ray" (the S half of T8N, R6E), and what had been BLACK EARTH (the N half of the same township) became part of MAZOMANIE, then newly organized. Thus the name "Black Earth" has applied to 3 different areas since its first use as a town name in 1851.

Black Earth, vill.

For the PO, and the creek on which it was platted, Aug. 1850, chiefly on sec. 26, BLACK EARTH, under direction of Ori(e)n B. Haseltine, of Vermont, and James T. Peck. Incorporated 1857.

Black Earth Creek (formerly also, —— **River**)

For the dark soil of its valley. Whether the name was original with the whites or translated from an Indian word is unknown; the latter is a good possibility, however, for the Indians gave many names descriptive of soil, topography, etc. (cp. Milwaukee, Okee, etc.).

"Black Earth River" is first shown on Finley's map of 1826, but too far N along the Wisconsin River; Mitchell (1835) follows suit, placing it above "Prairie des Sacs," and putting "Cheepee R." about where Black Earth Cr. should be. On Judson's map (1838) "Black Earth Cr." appears in about the right location, and so in later maps. It rises in NW MIDDLETON and flows W thru CROSS PLAINS, BERRY, BLACK EARTH and MAZOMANIE, into Iowa Co., where it joins Mounds Creek. Its tributaries have sometimes been shown on maps by its name —e.g., *Vermont Cr.;* the name "Duck Cr." (belonging in Columbia Co.) is placed on it on Chan., 1829.

Black Earth PO

For the creek. Est. Apr. 26, 1850; Ori(e)n B. Haseltine was postmaster, and it was kept on his place, sec. 22; it was later moved into the village.

Black Earth Valley

For *Black Earth Creek*, which runs thru it: a large, fertile valley stretching from MIDDLETON W to the Wisconsin River. First settlement was made in 1843. (Butt. 888)

Blackhawk, subd.

For chief Black Hawk, who passed here in his flight toward the Wisconsin R.; see next. Platted 1890; since absorbed into *Shorewood Hills.*

Blackhawk's Cave

For the Indian leader of the "Black Hawk War," who is said to have hidden here (1832) in his flight from L. Koshkonong to the *Battle Ground:* a small cave, accessible only from the water, in a cliff on the S shore of L. Mendota; SW sec. 17, MADISON.

Black Woods, The ['blæk'wudz]

Because of their thickness and darkness, in settlement days: a large wooded area, formerly covering much of SW WESTPORT, but now much reduced. (Corc.)

Blackwoods School (also, **Black Woods ——**) ['blæk,wudz]

For the *Black Woods:* on the town road in SW sec. 29, WESTPORT.

Blooming Grove, town

Said to have been a descriptive name suggested by the Rev. J. G. Miller, who "had never seen a section of the country in which there were such fine groves and so many wild flowers" (Keyes 1, 320). The name already existed, however, in N. Y., Tenn., and Ind. before 1842; so it may be in part reminiscent. The town was established Jan. 11, 1850, by separation from MADISON; it then included T7N, R10E, except secs. 5, 6 and 7. These were acquired not long after, bringing it to its present area.

Blooming Grove, vill.

For the town. The village was platted 1915; SE sec. 17, BLOOMING GROVE; since absorbed into *Monona Village.*

*** Blooming Grove, PO**

For the town. Est. Dec. 10, 1866; disct. Nov. 7, 1871. (Butt., 929, says a PO was est. in the town before 1850, but this must be an error.)

Blue Beach Park

An error on some maps for *Blue Bill Park.*

Blue Bill Park, subd.

For *Blue Bill Point,* which was included in it; on the Yahara R., platted 1909, in SE sec. 22 and NE sec. 27, WESTPORT.

Blue Bill Point

From the fact that blue-bill (=scaup) ducks were plentiful here; so called by hunters as early as 1865 (Corc.): a point of land on the right bank of the Yahara R. where it enters L. Mendota: NE sec. 27, WESTPORT.

Blue Bird Park

An error on some maps for *Blue Bill Park.*

*** Blue Mound PO**

For the nearby eastern "Blue Mound," at whose foot it was. See *Blue Mounds PO.*

*** Blue Mound Precinct**

For its location at *Brigham's:* an election precinct est. Apr. 3, 1843, including the present PERRY, BLUE MOUNDS, VERMONT, BLACK EARTH, and MAZOMANIE. It superseded the *Moundville Precinct.*

Blue Mounds, (The) [ˌblu 'maʊndz]

Descriptive of their appearance when seen at a distance; cp. the earlier names *Smoky Mountains, Old Smoky.* There are two hills, joining at the base, East Blue Mound being in Dane Co., West Blue Mound in Iowa Co. The latter is the third highest point in Wis. (1716 ft.). Visible for many miles in every direction, these "mounds" were the most prominent early landmark in the region, and are mentioned by many early travelers (for example, Jonathan Carver, tho he does not name them).

Their first notice on a map is on that of Arrowsmith, London, 1796, "A High Mountain"—presumably the western mound. From this comes "High Mountain" on the maps of J. Melish, Phila., 1814, and Carey and Lea, Phila., 1823. H. Huebbe, Gotha 1825, shows "Smocky Mts."—the first recognition of there being more than one. Chandler, 1829, is first to mark "Blue mound," and Farmer, 1830, "Blue Mounds." The definite article is sometimes used of the hills, never of the village or town which have adopted the name.

Blue Mounds, town

For *The Blue Mounds,* most prominent local geographical feature. When created (Mar. 11, 1848), the town included T6 &7N,R6E. The second of these townships was separated Nov. 16, 1855 as VERMONT, reducing BLUE MOUNDS to its present limits.

Blue Mounds, vill.

For the *Blue Mounds.* J. R. Brigham wrote, "The settlement . . . has always borne the name of Blue Mounds" (Park 241). An early settlement is shown here as *Moundville,* however, and *Beaumont* was platted in 1836. The village was not platted under the name of *Blue Mounds* till 1881 (SE quarter of SW sec. 6). It was incorporated 1912, and now includes parts of secs. 6 and 7, BLUE MOUNDS.

Blue Mounds Branch (1)

Because its sources are in BLUE MOUNDS: a branch of the *Pecatonica R.*, itself having two main branches, which rise in the W part of the town, meet in sec. 6, PERRY, and flow into Iowa Co. The name became official by action of the Wisconsin Geographic Board, June 11, 1941, displacing *East Blue Mounds Branch* and *East Branch*, former alternate names. Tanner's map (1833) labels this "Peektano."

* **Blue Mounds Branch** (2)

For its source near the Blue Mounds: an early name (1844: Dane 90) for the *East Branch of Mounds Creek*. See next.

Blue Mound(s) Creek

=*East Branch* (*of Mounds Creek*); this name has been used from at least 1849 (Lap.), and is still on Hiway maps.

Blue Mounds Fort

For the nearby Blue Mounds: a fort, built April, 1832, by Ebenezer Brigham and others, in SE sec. 7, BLUE MOUNDS, covering about one-fourth of an acre. Its site is now marked as a historical monument.

Blue Mounds PO

For the nearby mounds, ultimately; the plural form of the name, however, is significant, as this PO supplanted two others: *Blue Mound PO* and *West Blue Mound PO*. The history of these PO's and their antecedents, being confusing, is here given in detail.

Before any PO was officially established, Ebenezer Brigham, first white settler of Dane Co., handled what mail there was—from about 1828 till 1837, when *Moundville PO* was established, with J. C. Kellogg as postmaster. This was at the Brigham settlement, secs. 5, 7, and 8, BLUE MOUNDS.

Moundville PO was renamed *Beaumont PO* May 18, 1839; and this was renamed *Blue Mound PO* Feb. 9, 1843. When a new PO (*West Blue Mound PO*) was est. Mar. 6, 1857, just over the line in Iowa Co., at the foot of the West Blue Mound, *Blue Mound PO* was moved farther east, to the crossroads at the corners of secs. 10, 11, 14, and 15, and Thomas Haney, who owned the land here, became postmaster.

Blue Mound PO was disct. Nov. 13, 1866, re-est. Aug. 18, 1868; and finally disct. (as was also *West Blue Mound PO*) Dec. 8, 1881, when the two were superseded by *Blue Mounds PO*, in the present village of Blue Mounds.

Blue Mounds Road, The (1)

For the intended terminus, the Blue Mounds, which, however, it never reached. This (old state highway 30) was the first road from Milwaukee to Madison, and from 1837–9, the only one. It is still so called in Milwaukee (Eldred). Also, *Cottage Grove Road*, and early, *The Milwaukee-Madison Rd.*

Blue Mounds Road, The (2)

Because it joins Madison with Blue Mounds: another name for the *Mineral Point Rd.* (Barton 740, etc.)

Blue Valley

Supposedly descriptive, tho perhaps influenced by the proximity and importance of the *Blue Mounds*. The valley of a branch of the Little Sugar R., chiefly that part in which Blue Valley School and the Blue Valley Cheese Factory are: sec. 35, BLUE MOUNDS. So called from at least the 1890's; the name may not have come into being until the cheese factory had to be named. (Goebel)

Blue Valley School

For the valley: on the town road in S sec. 35, BLUE MOUNDS.

*** Bluff PO**

For the nearby *Pine Bluff*. Est. May 23, 1882, and kept at the then village of *Pine Bluff Station;* name changed Apr. 27, 1891, to *Klevenville PO.*

Blum's Creek [blumz]

For John Blum and his descendants, across whose land (SW sec. 18, ROXBURY and NE sec. 13, in the N part of MAZOMANIE) it runs. Only the lower part of the creek is so known; the upper part is called *Madison Cr.;* formerly *Inama's Creek* (Derl.) Blum acquired the land in the 1890's.

*** Boiling Creek**

Descriptive of the action of its water? A former name for the present *Dunlap Creek.* On maps from 1849 (Lapham) to 1855; also Stat.

Bohn School [bon]

For Ernst Bohn and his descendants, owners of the adjoining land from before 1873; the school is in SE sec. 28, VERMONT.

Booth School [buθ]

For George Booth, of Yorkshire, England (who settled on adjoining land, sec. 6, in 1860), and his descendants; the school is in NE sec. 7, VERMONT.

Borchers Beach, subd. (also, erron., **Borcher's** ——) ['bɔrtʃɚz] also (Corc.) ['bɝdʒɚz]

For William Borchers, owner of the land; platted 1907 on L. Mendota, in NE sec. 33, WESTPORT.

Borghilda Spring ['bo-rg,hɪldɑ]

For Borghilda Groven who discovered it about 1844 on the land on which she and her husband, settlers from Norway, had just squatted. It is almost exactly in the center of sec. 20, DEERFIELD, and feeds Koshkonong Cr. (Reque)

* **Bottolf Spring** ['bɒt,ɒlf]

For Bottolf J. Grinde, early Norwegian settler (1848). It is in NE sec. 30, DEERFIELD; it is now nearly dry. (Thor.)

Bowers School ['bɑʊɚz]

For Bower Bowers, born here in 1851, son of Norwegian settlers; the family homestead and the school are at the road junction, in extreme SE sec. 23, PRIMROSE.

Box Elder Grove School

For its location among box elder trees. At the junction in NE sec. 23, MEDINA.

Box Elder School

Descriptive of the surrounding trees. At the crossroads between secs. 22 and 23, BRISTOL.

* **Brackenwagen's,** cor.

For the storekeeper and postmaster, F. Brackenwagen, at the crossroads settlement, corner of secs. 10, 11, 14, 15, BLUE MOUNDS; so called in the 1870's. Now *Luhman's Corners.* (H&W 1873) Folk etymology makes this sometimes "Broken-Wagon." (Gilb.)

Brag Hollow, mine ['bræg ,hɑlɚ]

Because of the bragging done by two miners, Pat Sweeney and Hiram Smith, who struck a rich deposit of lead here about 1865: a ravine or hollow somewhat SW of the Cave of the Mounds, in NW sec. 8, BLUE MOUNDS. The lead was close to the surface, in a knoll, and it is said each man got two or three thousand dollars of lead out of it (MacK). The name is still well known locally, tho the location of the hollow is not. The pronunciation is traditional; Park spells the word "holler" (245).

Braun School ['braʊn]

For the Braun (or Brown) family, whose members owned land in sec. 18, CROSS PLAINS, from before 1890, but not next to the school till after 1911. A former name for *Garfoot School.*

Brazee's Lake (also, **L. Brazee**) [brə'ziz]

For David and Thomas Brazee, who settled on its E side in 1840: a lake on secs. 33 and (chiefly) 34, BRISTOL. The amount of water varies greatly from year to year; it was formerly of considerable extent and the water clear, but in dry seasons almost disappears. The USGS map was made in a dry time (1905) and therefore calls it Brazee Swamp; the surface is at present almost wholly covered with weeds; yet it is everywhere known locally as a lake, not a swamp. This name, tho considered the official one, is not widely known; other names it has borne are *Patrick's L.*, *L. Washington*, the *Old Lake*, and *Duscheck's Lake.*

*** Brazee Swamp**

See *Brazee's Lake.*

Break Neck Hill

From an accident fatal to J. B. Runey, first settler of OREGON, whose wagon overturned here (sec. 1, FITCHBURG) in Sept. 1846 (Butt. 1248). The name is still well known, tho little used.

Brecken Hill ['brɛkɛn]

For Andrew Brecken, owner of the land since before 1926: a hill in NW sec. 23, PLEASANT SPRINGS. (Juve)

*** Brereton PO** ['brɪrtən]

For Hugh H. Brereton, nearby resident, who was instrumental in having it established (Brere.). It was kept at the crossroads E of the center of sec. 30, DANE. Est. Mar. 7, 1872; disct. July 6, 1877.

Breunig's Hill ['brɔɪˌnɪgz]

For G. Breunig, German settler, whose family was in the town before 1890, but who did not buy land including the hill till between 1911 and 1926: a considerable hill, mostly in W sec. 17, ROXBURY. (Derl.)

Brewery Creek, The ['brurɪ]

Because of the brewery by which it flowed before entering Black Earth Cr.: a small creek which rises in the SE of BERRY, and flows into the east end of the vill. of Cross Plains. So called locally only. (Dahmen)

Briar Hill, subd.

Descriptive: it was on a slope thickly grown with berry bushes (Woodw.); platted 1916, in W sec. 28, MADISON.

Brickson Park, subd. ['briksn̩]

For Henry Brickson, owner of the land; on the Yahara R., platted 1909 in SE sec. 22 and NE sec. 27, WESTPORT.

Brictson Hill ['briksn̩]

For Edward Brictson (or Brickson), who settled here about 1848, and his family, still in possession of the land: a hill mostly in SE sec. 6, PLEASANT SPRINGS.

Brictson Park, subd. (also, erron., **Brickson**) ['brɪksn̩]

For former owners of the land, the last of whom was Emma J. Brictson; on L. Waubesa, platted 1918, in S sec. 8 and N sec. 17, DUNN.

* **Brigham Lead, The,** mine ['brɪgəm 'lid]

For Ebenezer Brigham, discoverer and owner: a lead mine opened in 1828 on sec. 7, BLUE MOUNDS. (Park 236)

* **Brigham's**

=next. So on Chandler's map, 1829, and other early maps. See also *Brighamsville.*

* **Brigham's Place (—— Tavern, Brigham Place)**

For its builder and owner, Ebenezer Brigham, who settled 1828. This was a widely known tavern, on sec. 6, close to the mines on sec. 7, BLUE MOUNDS, a center of the miners' life and beginning of the Blue Mounds Settlement. The site was still so called in 1877 (Park 236).

* **Brighamsville**

For *Brigham's;* only on Taylor's maps, 1838; sec. 6, BLUE MOUNDS.

Bristol, town

For Bristol, Ontario Co., N. Y., the former home of David Wilder, who suggested the name. Wilder settled here in 1842. The name was given when the town (T9N, R11E) was separated from SUN PRAIRIE, Mar. 11, 1848.

Bristol Church, The

For its location (NE sec. 29, BRISTOL): a Methodist Episcopal church, built 1865. (Park 384, Butt. 885)

Brittingham Bay ['brɪtɪŋ,hæm]

For the adjoining Brittingham Park (for Thos. E. Brittingham of Madison, donor): a small bay in the W end of L. Monona; sec. 23, MADISON. The park was made before 1911.

Britts Valley (also, **Britt** ——) [brɪts]

For George Britts, of Virginia, who, some time after 1852, built a flour mill here: the upper valley of the W branch of Sugar R., especially the part in sec. 6, PRIMROSE, where the mill was. The mill burned about 1885 (Austin), but the name is still in use.

Britts Valley School

For *Britts Valley*, in which it is: S sec. 6, PRIMROSE.

Brockton, subd. ['brɑktən]

For Brockton, Ill., thru which Ross M. Koen had frequently passed, and the pleasant name of which had attracted him; within the city of Madison, platted 1927; NE sec. 6, MADISON.

Brooklyn, vill.

For BROOKLYN, Green Co., in which it mostly is (tho additions have been made in Dane Co., W sec. 31, RUTLAND). When the C and NW railroad was coming thru Green Co., H. B. Capwell platted this village, and became the Station Agent (1864). The Railroad wanted to name it Capwell, but he preferred to have it Brooklyn. (BROOKLYN was so named by a committee of New Yorkers for the city in their native state.)

Brookside, subd.

For its position beside Starkweather Creek, which, however, has never been called a "brook." ("Brook" is evidently used for its connotation, as being less common than "creek.") Probably also reminiscent. Platted 1924, within the city of Madison.

Bryngelson's Hill ['brɪŋgl̩sənz]

For Sever Bryngelson, Norwegian settler, owner of the land from before 1873 till after 1904: a small hill on the edge of L. Waubesa, in E sec. 4, DUNN.

Bryn Mawr Church [ˌbrɪn'mɑr, —— 'mɔr]

For Bryn Mawr, Pa., from which place the financial support originally came: a Presbyterian Church, built in 1896, in Cottage Grove.

* **Buckeye,** cor.

For the *Buckeye Tavern* here: an early crossroads settlement. *Door Creek PO* was est. here (1847), and became an alternate name, but Buckeye is still remembered locally. (Thor., etc.)

Buckeye School

For *Buckeye* corner, where it is.

*** Buckeye Tavern**

Said to have been run by people from Ohio, the "Buckeye State" : an important early tavern (1840's and after) in NE quarter of the SE quarter, sec. 33, COTTAGE GROVE, just E of the juncture of two Indian trails, which became the *Jefferson and Madison Road* and the *White Water and Madison Road.* (Lig.)

Bullhead Lake ['bʊlɛd]

Because it was once stocked with the fish so called, for which fishermen came from as far away as Chicago (Lein): local unofficial name for *Rice L.*

Burgess' Corners ['bɝdʒəs]

For the Burgess brothers, who owned adjacent land from before 1911 till after 1931: the crossroads between secs. 3, 4, 9 and 10, SUN PRAIRIE.

Burke, town

For the Irish statesman, Edmund Burke (d. 1797), who defended the American colonies before the Revolutionary war. Tho settlement began here in 1837, the town (T8N, R10E) was not established till Nov. 18, 1851 (by separation from WINDSOR). Many Irish were among the first settlers, and probably influenced the choice of the name.

Burke, vill.

For the town: platted as a village in 1886, in S sec. 23, but the settlement existed much earlier. See *Burke Station.*

*** Burke Center PO**

For its approximate location in the town. Est. Apr. 11, 1863, and probably kept at the Burke Cottage Tavern, NE sec. 14, which was run by the postmaster, Adam Smith. Disct. Sept. 20, 1869.

Burke Creek

For BURKE, in which it rises (sec. 29): it flows S into sec. 5, BLOOMING GROVE, to join Starkweather Cr. (It is named only on the plat of Fair Oaks and its additions, 1905.)

*** Burke PO**

For the town. Est. Sept. 28, 1854, and seems to have been kept first on sec. 14, then sec. 11, until the establishment of *Burke Center PO* (1863), soon after which it was discontinued, Sept. 7, 1864. Burke Center PO ran only till 1869; Burke PO was reest. July 28, 1886, in the village, where it remained till its final discontinuation in 1920. See *Berk PO.*

Burke Station

For the town: another name for the village of *Burke*, with reference to the railroad station there, discontinued about 10 years ago; the name is still used, however.

Burke Station School

For nearby *Burke Station.* In NE sec. 26, BURKE.

* **Burnson Lake** ['bɜnsn̩]

For B. H. Burnson (originally Björnson), a Norwegian settler, who owned the land bordering it from before 1873 till after 1911. It was drained several years ago. Another name for *Norwegian L.* (Renk)

Burritt School ['bɝit, 'bɝt]

For Frank Burritt, who owned adjacent land: SW sec. 32, PLEASANT SPRINGS. (Olson)

Busseyville Creek, The ['bʌsɪvɪl]

For the village of Busseyville (SW Jefferson Co.), thru which it passes: an unofficial local name in Dane, Jefferson, and Rock counties, in the region of L. Koshkonong, for *Koshkonong Cr.*

Buss's Corners ['bʌsɪz]

For Fred Buss and his family, owners of the adjoining land since before 1899: a crossroads at the juncture of secs. 8, 9, 16, and 17, SUN PRAIRIE.

Butler's Point

For J. Butler, owner since the early 1930's: a piece of land jutting northward into L. Waubesa; N sec. 9, DUNN.

Byrne's Hill ['bɜnz]

For the Byrne family (Lawrence Byrne bought the land before 1890): the N edge of *White Oak Hill*, in N sec. 26, FITCHBURG. (Barry)

Byron Long's Hill

For the owner of the land to the SW of it from some time between 1873 and 1899: an alternate name (no longer well known) for *Hippe's Hill.*

* **Camanche,** vill.

Probably for the Indians of this name (later spelt "Comanche"). These Indians, native to Texas, had become known in the first decade of the 19th century, and the name was being applied (as here) where it did not belong. It is found as the name of an Indian village on maps of 1838 (Mitch., Hinman, Taylor), at the extreme S edge of Dane Co., and W of Sugar R. Mitch. puts it in SW MONTROSE, the others in SE PRIMROSE (thru which an Indian trail ran). The application of this name to this spot must have been the doing of some imaginative white man. It is perhaps noteworthy that Dr. George Peck had, in 1836, used this name for the city he was promoting in Clinton Co., Iowa, where it was no more native than here. It may well be that there is some connection between these two uses of the name, only 2 years apart. Also called *Chemanchville.*

Cambridge, vill. [ˈkemˌbrɪdʒ]

For Cambridge, Washington Co., N. Y. Said to have been given by Alvin B. Carpenter (one of the platters) to commemorate the residence of a boyhood sweetheart (Dow; MCT, 9.19.41). Platted in 1847, in NE sec. 12, CHRISTIANA.

Cambridge PO

For the village; est. here July 25, 1848.

Camp Badger School

For the nearby Camp Badger shooting range of the State Militia, used about 1910–6. In SW sec. 5, FITCHBURG.

Campbell Hill [ˈkæml̩]

For Edward, Hugh, and John Campbell, of Virginia, who in 1839 settled near Pine Bluff; they built a stone house which became a stopping place for travelers. The hill is on the W part of secs. 24 and 25, CROSS PLAINS. (Barton, Dahmen)

Campbell's Hill

For Thomas Campbell, who owned the land: a hill mostly in SE sec. 18, WINDSOR. (Linde)

Camp Brooklyn, subd.

For the nearby village of Brooklyn (Green and Dane Counties), from which its promoters, J. H. Richards and G. I. Tripp, came; on L. Kegonsa, platted 1898, in SE sec. 25, DUNN. Many Brooklyn people bought cottages here, making the name all the more appropriate.

Camp Columbia, subd.

For patriotic reasons? On L. Kegonsa, platted 1893, in SE sec. 25, DUNN.

Camp Dewey, subd. [ˈduɪ]

For Admiral Geo. Dewey, who had just won the battle of Manila (May 1, 1898); on L. Kegonsa, platted July 25, 1898, by Thore Olson, in SE sec. 25, DUNN.

Camp Gallistela [ˌgælɪsˈtɛlə]

For Mr. and Mrs. A. F. Gallistel, who were instrumental in founding and making it successful: the "tent colony" summer camp of the University of Wisconsin, on L. Mendota; S sec. 9, MADISON. The name was suggested by Dean Scott Goodnight, and chosen by vote of the students at the camp, about 1925. (Gal.)

Camp Leonard, subd. [ˈlɛnɚd]

For Leonard Larsen, their son; so named by Mr. and Mrs. Hans C. Larsen, owners; on L. Waubesa, platted 1905, in SE sec. 4, DUNN.

Camp Randall [ˈrændəl]

For Gov. Alexander W. Randall, when, in 1861, it was put into use as a training-ground for troops. In 1893 it was purchased by the University of Wisconsin for an athletic field. NE sec. 22, MADISON.

Camp Sunrise, subd.

From its position (on the NW shore of L. Mendota) from which the sunrise may be seen across the lake; platted 1913, in SW sec. 6, MADISON.

Capital City View, subd. [ˈkæpətəl]

Because from it a view of Madison, capital of Wisconsin, may be had; platted 1926, in SW sec. 17, BLOOMING GROVE.

Carpenter School

For Thomas L. Carpenter and his family, who have held adjacent land to the N since before 1861. The school is in NW sec. 34, CHRISTIANA. Now closed.

Casey's Corners

For Michael Casey, who owned the adjoining land from before 1873 until after 1911: the crossroads between secs. 27 and 34, CROSS PLAINS. (Dahmen)

Castle Place, subd.

For a castle-like house which had been built nearby (1863) by Benjamin Walker (Barton); within the city of Madison, platted 1903.

Catfish [ˈkætˌfɪʃ]

For the *Catfish Creek:* the point at which it is crossed by state highway 113 (present "Catfish Bridge") and the C&NW railroad, between secs. 22 and 23, WESTPORT. (USGS map, 1904, Madison Quadr.)

*** Catfish Bridge**

For the creek: the first bridge to cross it in the Madison area. Built in 1841 at *Catfish Ford* (BURKE). (Dane) This was not the present "Catfish Bridge"; see preceding entry, and *Black Bridge*.

Catfish Creek or **River**

A former name of the *Yahara R.*, officially superseded since 1855, but still in widespread use. The word is probably a translation from Ojibwa (see *Badfish Cr.*). "Catfish Cr." first appears on the MS. map in the surveyor's notebook for T5N, R11E, 1833, and thereafter on other early maps. "Catfish R." first appears on the I&W map, 1838. Maps continued to give Catfish as an alternative of Yahara at least as late as 1926. "Catfish," in early use in Madison, meant specifically that part of the stream between lakes Mendota and Monona (1845: **Dane** 118 ff.).

*** Catfish Ford** (BURKE)

A ford of the Catfish R. Black Hawk and his followers crossed it in their flight in 1832. It was where the Williamson St. Bridge, Madison, now crosses the stream. (Park 16; WHS *Coll.* IV, 345)

*** Catfish Ford** (WESTPORT)

A ford of the Catfish R. used by the Winnebago Indians. Orson Lyon, in his surveyor's notebook (1834), shows it without name near the S edge of sec. 27; but when the level of L. Mendota was raised (1858) it was submerged, and the crossing was moved to SE sec. 22: the present *Catfish*.

Catfish School

For the *Catfish Creek*, which passes near by. Between N secs. 22 and 23, WESTPORT.

*** Cat Island**

Origin unknown. A former name of *Rocky Roost*. (Sull.)

Cave of the Mounds

For the nearby Blue Mounds: a sizable cave, discovered 1939, in SW sec. 5 and NW sec. 8, BLUE MOUNDS, on the land of C. I. Brigham. (MCT, 9.14.41)

Cedar Bay, subd.

Descriptive (there is a small bay here in the L. Mendota shoreline, with a fringe of cedars on the bank); platted 1908, in SE sec. 6, MADISON; since, mostly absorbed into *West Point*.

The name was suggested by Mrs. Sophie M. Briggs, who lived nearby. (Briggs)

Cedar Bluff

Because it is the only one in the locality which is covered with cedars: a large bluff mostly in SE sec. 9, DANE. (Steele)

Cedar Edge School

For some nearby cedars. On highway 73, between secs. 9 and 10, YORK. Now closed.

*** Cedar Point**

For the cedar trees growing there: an early name for the present *Lund's Point.*

Center Bluff

Because of its position between branches of Spring Creek and the ranges of bluffs to north and south: a striking isolated bluff mostly in NW sec. 8, DANE. An early name, still fully current.

Central Home Addition, subd.

Because centrally located, and intended for residential purposes: within the city of Madison, platted 1914.

Channel Cat, The

For its shape, with a large head, like a catfish: a small island in the channel between Railroad Bridge Is. and the Dane Co. mainland. Known by this name to fishermen. (Derl.) Sec. 13, N part of MAZOMANIE.

*** Chase School**

For D. S. Chase and family, who settled before 1861 on land adjacent to the school. Also known as *Ames School* and *Babbitt School.*

*** Cheekee River**

Origin and meaning unknown. Probably an error for *Cheepee R.* The name "Cheekee" is placed on what is today the *East Branch of Mounds Creek* by Tanner, on his maps of 1839 and 1845.

*** Cheepee River**

Origin and meaning uncertain. Possibly an attempt to render Ojibwa ['tʃiˌpaɪ] meaning "corpse" (Geary, Lincn.). It is found only on Finley's (1826) and Mitchell's (1835) maps, placed on a stream about where the present *Black Earth Creek* should be; but since, on these maps, Black Earth Creek is itself placed too far north along the Wisconsin R., "Cheepee R." should probably be identified with *Cheekee R.*, as an early name for the E Branch of Mounds Creek.

*** Chemanchville**

=*Camanche.* (The Taylor map uses the form "Chemanchville.")

Cherokee Marsh

For the Cherokee Hunting Club, whose clubhouse is here. The club was est. about 1887 (Corsc., Fraut.), tho why it was so named is not clear. The marsh is mostly in the E part of WESTPORT.

Chicago and Northwestern Railroad

For the region it serves; formerly, in Dane Co., the *Beloit and Madison RR* and the *Baraboo Air Line*, the two running N and S thru the county. There is also a line running E and W, the E passing thru Cottage Grove and Deerfield and leaving the county from London (completed 1881); the W passing thru Verona, Mt. Horeb, and Blue Mounds, and leaving the county from sec. 6, BLUE MOUNDS.

Chicago, Milwaukee, St. Paul and Pacific Railroad

For the major places on its routes; formerly the *Milwaukee and Mississippi RR;* built thru Dane Co. from sec. 32, ALBION, NW thru Madison to sec. 18, MAZOMANIE. It reached Stoughton in 1853, Madison in 1854, Mazomanie in 1856. There is also a branch N from Mazomanie to Sauk City, built about 1881. See also the *Madison, Sun Prairie, and Watertown RR*, and *Madison and Portage RR.*

Chicken Hill

Because everybody on it seemed to raise chickens: a small hill in the N part of sec. 15, DUNKIRK; a local, humorous name. (Johns.)

Christiana, town [ˌkrɪstɪ'ænə]

For the PO: T6N, R12E, formed by separation from ALBION, Jan. 3, 1847.

*** Christiana PO**

For the capital of Norway. Nilsen's account is generally accepted: Gunnul Olson Vindæg, a Norwegian who had settled here in 1840 and become prominent, chose the name, "but as he was not particularly skilled in spelling, he unfortunately happened to write Christiana instead of Christiania, and this designation has since been retained."

The PO was est. Feb. 2, 1846, the postmaster being Wm. M. Mayhew (see *Mayhew's*), who kept it on sec. 28 till the fall of

1849, when it was moved just south of the village of Clinton. Efforts to change its name to Clinton PO failed because there was a PO already by that name; however, on Dec. 9, 1885, it was changed to *Rockdale PO*.

*** Christina,** vill. [ˌkrɪsˈtinə]

For his wife, by Peter L. Mohr, who platted it Oct. 25, 1855, in NE sec. 3, CROSS PLAINS. It was the second of the 3 villages which, in part or whole, were united (1920) into the present *Cross Plains*. It was known locally as "middle town."

*** Cincinnati Heights**

For Cincinnati, Ohio, former residence of J. M. Dickson, the owner (1877; Park 415–6). The name was properly that of the farm as a whole, but also seems to have been used specifically of the feature which had furnished its former name, *Rock Terrace*.

*** City of the First Lake**

For its location on First L. (now L. Kegonsa): a "paper town" in sec. 23, DUNN. Shown on early maps (tho not always in the same position) from 1836 (Judson) forward. No plat was recorded, and it was never settled as such.

*** City of the Four Lakes**

From its location at the head of *the Four Lakes*. A "paper town," projected along the NW shore of L. Mendota, sec. 6, MADISON, on the left side of Pheasant Branch Creek. The plat was recorded July 22, 1836; the promoters were Morgan L. Martin, Col. W. B. Slaughter, and J. D. Doty, and it was proposed as the new capital of the state, in competition with Madison (itself unbuilt) and other towns; but the entire project failed. It is shown on early maps from 1836; it was considered officially defunct by 1843.

*** City of the Four Lakes PO**

For the *City of the Four Lakes*. Est. Feb. 16, 1837, disct. Aug. 9, 1837; re-est. Nov. 14, 1839, disct. Feb. 25, 1841.

*** City of the Second Lake**

For its location on Second L. (now L. Waubesa), on both sides of the Yahara R. where it flows out of the lake: a "paper town," platted Oct. 11, 1836, in sec. 3, DUNN. It never was settled as such, but the land it covered is now within the village of Macfarland.

*** Clark's Corners**

For John Clark, who settled here before the middle of 1845: an early name for *Springfield Corners.* How long the first name lasted, and when the second began to be used is uncertain. Clark was postmaster here (Dane PO) till at least 1849; this was changed to Springfield Corner PO Nov. 24, 1871.

*** Clark's Lake**

For John T. Clark, owner in the 1870's of the land bordering it, secs. 3 and 4, ROXBURY. This name was concurrent for a time with *Fish Lake,* its present name.

*** Clarkson,** town

Of uncertain origin: a former town, est. Feb. 11, 1847, including T9N,R7&8E (the present ROXBURY and DANE); the name was changed to *Dane,* Mar. 11, 1848. The first town meeting as held at the house of John Clark (probably at *Clark's Corners*) —and the town may have taken its name in part from his.

*** Clarkson PO**

For James S. Clarkson, First Asst. Postmaster General at the time. Est. Nov. 29, 1889, and kept on his land (NW sec. 14, PLEASANT SPRINGS) by Aslak K. Juve, the first postmaster. It is said the name was to have been "Juve," but since there were others in the state too similar to this, Juve suggested "Clarkson" (Juve). Disct. Nov. 30, 1901.

Clay Ridge School

Descriptive of a local feature. Where secs. 5, 6, 7, and 8 meet, YORK.

*** Cleveland's Cave**

For Benj. Cleveland, owner of the land from before 1847; shown on maps of 1849 and 1855 (Lapham, Gree.). See *Richardson('s) Cave.*

Clifton, vill.

Partly descriptive, and probably reminiscent: a village, platted 1851, in NW sec. 6, ROXBURY, intended to be on and below the cliffs (i.e., bluffs) of the Wisconsin R. It never grew as large as had been planned; the name is still known, but not much used.

*** Clinton,** vill.

Probably for Clinton, Oneida Co., N. Y.: a "paper town" at first, it was platted Nov. 17, 1836, but the first house was not built till 1846. However, maps show it, from Terr. forward; NE sec. 23 and NW sec. 24, CHRISTIANA. In 1885 it was replatted, and the name was changed to *Rockdale.*

*** Clinton Precinct**

For *Clinton:* an election precinct est. Apr. 3, 1843, and including the present DUNKIRK, PLEASANT SPRINGS, ALBION, CHRISTIANA, and DEERFIELD.

*** Clontarf,** vill. (also, **Clantorf**; see second pronunciation) ['klɑn-tɑrf, 'klæntɔrf]

For the PO. The store in which the PO was kept was opened in 1864, and became the center of this small settlement; the first use of the name seems to have been for the PO, however; and with the discontinuation of the PO, the name gradually died.

*** Clontarf PO**

For Clontarf, Dublin Co., Ireland, the PO being in an Irish settlement. It was first kept by Patrick Carr, at his store at the crossroads between secs. 11 and 12, SPRINGDALE. Est. May 1, 1867; disct. Dec. 26, 1883, soon after establishment of *Riley PO* at the railroad station nearby.

*** Clyde Creek**

Probably so named by Robt. Ogilvie for the Clyde R., Scotland. Ogilvie, a breeder of Clydesdale horses, owned the farm and spring from which this creek (now *Starkweather Cr.*) rises, sec. 27, BURKE (Park). The name apparently gained little currency. In use in 1877.

Coffeytown

=*Coffeyville;* a common variant.

Coffeyville, vill. ['kɑfiˌvɪl]

For the P. Coffey family, which settled before 1861 (Lig.) in N sec. 20, COTTAGE GROVE; there were several houses grouped together, forming a small community. The name was never official, but widely used locally.

Colby School ['kolbɪ]

For G. R. Colby, of N. Y., and his family, who came to OREGON in 1850, and settled before 1873 on the land on which the school is: NW sec. 18. Now closed.

Colladay Park, subd. ['kɑlədɪ]

For the Colladay family, and specifically William E. Colladay, owner of the land; on L. Kegonsa, platted 1926, in N sec. 23, DUNN. It adjoins parts of *Washington Park*.

Colladay's Point (also, **Colladay ——**)

For Wm. M. Colladay, of Philadelphia, who settled here 1844

(Barton): a point of land jutting into L. Kegonsa, SW sec. 23, DUNN. It was platted 1939 by Chas. M. Colladay.

*** Colladay School**

For the prominent early family (see *Colladay's Point*). A former name of *Lakeside School.*

*** College Hill**

As the site of the first buildings of the University of Wisconsin. One mile W of the Capitol, Madison. So named in Regent, 1849, and Stat., 1851, but superseded by *University Hill;* currently called by the students *Bascom Hill,* or simply *The Hill.*

College Hills, subd.

Presumably because it borders on the University of Wisconsin lands, and perhaps intended to attract University people (the streets are also named for colleges and universities); platted 1912, in SW sec. 16, MADISON. Since absorbed into *Shorewood Hills.*

College Park, subd.

So named to encourage settlement by people connected with the University? Platted 1908, in NW quarter, sec. 21, MADISON; since absorbed into the city of Madison.

Columbus Road, The

For Columbus, the first important town on it outside of Dane Co. It was authorized in 1838, surveyed 1840, built shortly after. Its early course from Sun Prairie took it N to almost the center of BRISTOL, then E to Waterloo Cr., whence NE in virtually the same course as present U. S. highway 151. (Eldred; Barton 1033)

*** Cos-ca-hó-e-nah River**

For discussion of the meaning, see *L. Koshkonong.* This was an early name for the *Yahara R.*, used only on Chandler's maps 1829. See also *Goosh-ke-hawn R.*

Cottage Grove, town

For the village and PO. The town, when est. Feb. 11, 1847, included T7N,R11&12E; separation of DEERFIELD (Mar. 17, 1849), however, reduced COTTAGE GROVE to its present area, T7N,R11E.

Cottage Grove, vill.

For a grove of burr oaks in which, in 1840, Wm. C. Wells, one of the earliest settlers, built a cottage. Here, in 1841, Amos Beecher opened the first hotel and tavern in the town, the "Cottage Grove House," where also the PO was kept, and around which the settlement grew (sec. 7, COTTAGE GROVE).

When the C&NW railroad was later built thru the town (1881–2), the Cottage Grove PO was moved to the N part of sec. 16, where the station was, and many Cottage Grove merchants also moved to the new settlement. The original one became Old Cottage Grove, then *Vilas;* the new one was platted 1882, incorp. 1923, and now includes parts of secs. 9 and 16.

Cottage Grove PO

For the settlement, at which it was kept. Est. May 8, 1841; disct. Mar. 30, 1855; re-est. April 18, 1855. It was moved in 1882 from Old Cottage Grove to the present village.

*** Cottage Grove Precinct**

For the settlement: an election precinct, est. here June 29, 1846. (Park 545)

Cottage Grove Road

For *Cottage Grove*, thru which it passed: the first road from Milwaukee to Madison. So called as early as 1844 (Dane 93); see *Blue Mounds Road* (1).

Cottage Park, subd.

Descriptive: a park-like area intended to have cottages; platted 1889, in SE sec. 30, PLEASANT SPRINGS.

Country Club Addition, subd.

In reference to a country club which was being planned at this time; within the village of Sun Prairie, platted 1918.

Countryside Addition, subd.

Descriptive; platted 1935, in NE sec. 13, MIDDLETON.

*** Cow Spring**

Because cattle watered here; a small spring near *Vilas Spring*, south of L. Wingra. (Brown, 7)

Coyle School [kɔɪl]

For Edward Coyle (who settled on adjoining land before 1861) and his family. Formerly in NE sec. 35, now in SW sec. 26, CROSS PLAINS. Now closed.

Cranberry Island

Because cranberries grew here but not elsewhere in the neighborhood (Quam): an alternate local name for *Timber Is.*

Crane Lake

For the presence of cranes? Or perhaps for Alvin Crane, an early settler in this region (1841: Dane). Shown without name on maps from 1837 (Mor.) to 1848, when it appears (Taylor) as *Eshunikede Lake;* first named Crane L. on map of 1850 (Lapham). Secs. 14 and 15, ROXBURY.

*** Crawfish River**

Descriptive; the creek now so called does not flow thru Dane Co., but *Waterloo Cr.*, a tributary, is shown as "Crawfish R." on Suydam's map, 1836.

Crescent Park, subd.

For the shape of the L. Waubesa shore along which it lies; NW sec. 9, DUNN. The name was given by E. N. Edwards, who, with his brother Theodore, had it platted, 1901. (Edw.)

Crestwood, subd.

Descriptive: there are woods on the crest behind the plat. The name was suggested by Mrs. E. P. McKinney. Platted 1938, in NW sec. 19, MADISON.

*** Crockertown,** vill. ['krakɚˌtaʊn]

For the Crocker family, of N. Y., one of the earliest in the town (J. R. came in 1842, J. H. and W. W. shortly after), who settled on adjoining lands in secs. 30 and 31, MONTROSE, and were influential in that part of the town (Barton). The name was local and never official.

*** Crockertown School**

For nearby *Crockertown:* a former name, current in 1911 (Cant.), for *Montrose School.*

Crossing School

For its location: at the corner where county highways I and K, and state 113, cross, WESTPORT.

Cross Plains ['krɔs'plenz, 'kras'plenz]

The name is usually said to be descriptive of the plains here where two early roads crossed: the military road from Fort Crawford (Prairie du Chien) to Fort Howard (Green Bay), surveyed 1832, and the road from Arena to Madison. The implication is that the name was first used of the topographic feature, and taken from that for the village, town, etc. But this is probably not the fact, as an investigation of the early records shows. The earliest use of the name "Cross Plains" in this locality is in the name of the PO, established 1838, with Berry Haney as postmaster. The name is shown on no map before 1840; maps which mark the location at all, or records which allude to it, call it *Steele's* or

Haney's. The second road is shown first clearly on Taylor's map of 1838, which first shows *Haney's*. Thus it is most unlikely that the name was in local use before Haney had the PO established bearing it, and the explanation about the roads crossing is probably ex post facto. See *Cross Plains PO*.

Once the name was given, however, its obvious appropriateness earned it quick currency. The "plains" thru which the roads ran are the valley of Black Earth Cr., and that of the Brewery Cr. and (to the S) highway P.

Cross Plains, town

For the PO. The town (T7N,R7E) was est. and so named Feb. 11, 1847.

Cross Plains, vill.

For the PO, est. here 1838. Tho the PO lapsed within the year, the name continued for the locality and the settlement which had begun. Parts of this settlement were platted as *Christina* (1855), and *Foxville* (1857); a third plat was named *Cross Plains* (1859) and included what was then locally called *Baerville*, in NW sec. 2. Parts of the three settlements were combined and incorporated as the present village of Cross Plains, 1920, including parts of secs. 2, 3, and 4, CROSS PLAINS.

Cross Plains PO

Probably for some town or PO so named elsewhere. There was a Cross Plains PO before 1822 in Fayette Co., Kentucky, and since Berry Haney, the first postmaster of Cross Plains PO, Wis., who undoubtedly gave the name, is known to have been in Kentucky before coming here, and since (see *Cross Plains*) the name does not seem to have been here before the establishment of the PO, the likelihood is that he brought the name and applied it to this new spot, where it was appropriate enough to the topography. The PO was est. Mar. 7, 1838, and kept by Haney; disct. Nov. 9, 1938; re-est. July 27, 1850, and kept at *Baertown* by Sylvester Morgan. It was later moved down into the main village. The PO name has thus been extended to the locality in general, the settlement which developed into the present village, a voting precinct, and the town.

Cross Plains Prairie

For the nearby *Cross Plains:* the prairie in the valley of Black Earth Cr., in the N part of CROSS PLAINS, and W of the village. (Butt. 875)

*** Cross Plains Precinct**

For the Cross Plains settlement, its center: an election precinct est. April 2, 1844, and including the present SPRINGDALE, CROSS PLAINS, BERRY, ROXBURY, BLUE MOUNDS, VERMONT, BLACK EARTH, and MAZOMANIE. (Dane)

*** Crossings**

Apparently a shortening of *Grand Crossings.* (Burr 1836)

Crown Point

For its shape, it being the E end of *Barber's Hill*, and therefore rising above the general shoreline: a point of land jutting into L. Kegonsa, in NE sec. 26, DUNN. It was platted as a subdivision, 1906, by Ida Barber.

Crystal Lake (also, formerly, **Cristal ——**, **Chrystal ——**)

For its transparency; the water, tho shallow, is clear: a lake in secs. 1 and 2, ROXBURY; also partly in Columbia Co. First named on Can. map, 1859.

Crystal Lake School

For nearby *Crystal Lake:* near the crossroads in SE sec. 6, DANE.

Crystal Springs Park, subd.

Descriptive: there are two springs close to the edge of L. Kegonsa, within this area; platted 1899, in S sec. 30, PLEASANT SPRINGS.

Daentl's Lake ['dɛntl̩z]

For Joseph Daentl and family, owners from before 1890 till after 1904: a small lake in SE sec. 8, SPRINGFIELD.

Dahl Cave [dɑ:l]

For O. E. Dahl and his descendants, owners of the land since before 1873: a small cave in SE sec. 19, SPRINGDALE.

Dale Heights, subd. [del]

For the maiden name, Dale, of the mother-in-law of Ernest N. Warner, who platted and named it, 1909; on high ground in W sec. 18, MADISON. (Woodw., etc.)

Daleyville, vill. (sometimes, **Dahleville**) ['delɪvɪl]

For Onon B. Daley (originally Dahle) and his descendants: a village in adjoining parts of SW sec. 8 and NW sec. 17, PERRY. Here, in 1853, Dahle, a native of Norway, opened the first store in the town, made himself unofficial postmaster for the neighborhood, bringing the mail from Blue Mounds, and in other ways fostered a settlement. The name was in use before it was put on the maps (Thrift, 1931; etc.) certainly by 1880, probably earlier. Not platted till 1944.

Daleyville School

For the village. In SW sec. 8, PERRY.

* **Daly School** ['delɪ]

For the Daly family (see *Daly's Hill*). A former name for *Hillside School*, DUNN.

Daly's Hill (formerly, **Daley's** ——)

For Knute H. Daley, who, before 1873, bought the land in N sec. 15, DUNN, on which part of the hill is; see also *Olson's Hill*. The hill itself is in secs. 14, 15, 22, and 23, and the town line road comes up sharply on it. Daley's present descendants spell the name "Daly."

Dane, town [den]

For the PO, and ultimately the county. When the name was first applied to the town (formerly *Clarkson*) Mar. 11, 1848, this included T9N,R7&8E. However, with the separation of ROXBURY (Mar. 21, 1849) Dane was reduced to its present area, T9N,R8E.

Dane, vill.

For the town: a settlement, platted 1872, in adjoining corners of secs. 13, 14, 23, and 24, DANE; incorporated 1899.

Dane County

In honor of Nathan Dane, of Massachusetts, framer of the Ordinance of 1787 for the government of the Northwest Territory, out of which Wisconsin was the last territory to be formed. The name was suggested by Judge (later Governor) James D. Doty to the Hon. Ebenezer Brigham, first white settler of the county (Noonan), and was bestowed 2 years before organization of the county (Keyes 35), which was formed out of Milwaukee and Iowa counties, and constituted during the first session of the Territorial Legislature, Dec. 7, 1836. Until 1849 it included all of T9N,R6E, but in that year the part W of the Wisconsin R. was taken away, which reduced Dane Co. to its present area. (Revised Statutes, Ch. 2.)

Dane PO

For the County. It was est. July 22, 1844, with Freedom Simons as postmaster; therefore probably kept on his land in sec. 4, SPRINGFIELD. In 1845 John Clark became postmaster, and the PO was moved to *Clark's Corners* (SE sec. 5, now Springfield Corners). It was disct. Feb. 21, 1856; re-est. Mar. 25, 1856. Meantime there had been formed SPRINGFIELD (1847) and DANE (1848), Dane PO being in SPRINGFIELD. This confusing situation continued for many years, until it was resolved by two changes of name: Dane PO to *Springfield Corners PO*, and *Harvey PO* to *Dane Station PO* in DANE, both Nov. 24, 1871. Thus Dane PO lapsed for a time, but on June 22, 1883, Dane Station PO was renamed Dane PO, which it has remained since.

*** Dane Precinct**

For *Dane PO:* an election precinct est. June 29, 1846. (Park 545) It was at *Clark's Corners*, where the PO was then kept. Not mentioned in Dane.

Dane Station

For the town, and the railroad station established here (NW corner of sec. 24) in 1871. The *Dane Station PO* took its name from this in the same year, and tho the village was platted as "Dane," it was frequently known as "Dane Station" because of the station and PO, and because there was a *Dane PO* elsewhere.

*** Dane Station PO**

For the railroad station newly made (1871) on sec. 24, DANE. The name was changed from Harvey PO, Nov. 24, 1871, and was itself changed to Dane PO, June 22, 1883.

*** Daniel Baxter Road, The**

For an early settler, probably in Green Co., tho he has not been identified: a road so called in the 1840's (Park 448). Its course is not now altogether clear, but it ran N and S thru E Green Co., entering Dane Co. apparently in SE OREGON, and joining the *Fitchburg Rd.*

*** Danish Church, The**

For the origin of the settlement in which it was: the Danish Evangelical Lutheran Church in NE sec. 28, RUTLAND; incorporated 1883. The building was razed about 1930.

Danz School [dɔnts]

For A. A. Danz, who has owned the land on which it is from before 1911. In extreme SE sec. 17, BERRY. Formerly *Spring Valley School.*

Darwin Station

Origin unknown: a station on the CMSP railroad, est. 1891 in NE sec. 30, BURKE. It is still listed by the RR guides, tho disused for several years.

Deaddog Hole

Origin unknown: a particularly deep spot in the E channel of the Wisconsin R. below the *Channel Cat.* So called by sportsmen. Tradition has it that a dead dog was found floating here about 1898. (Derl.)

*** Dead Lake**

A name (about 1840) for *Lake Wingra.*

Dead Lake Ridge

For *Dead Lake*, a former name of L. Wingra, which this ridge separates from L. Monona. Mostly NW corner, sec. 26, MADISON.

Deansville, vill. (also, **Deanville**) ['dinzˌvɪl]

For Richard Dean, who (according to Park, Butt., etc.) platted it in 1860: a village in secs. 5 and 8, MEDINA. (No plat has been found in the Dane Co. courthouse.)

*** Deansville PO**

For Richard Dean, first postmaster. Est. May 4, 1860; disct. Jan. 15, 1900.

Deansville School

For nearby *Deansville.* In SW sec. 8, MEDINA. (Dusch.)

Deansville Station

= *Deansville;* there has been a station of the CMSP Railroad here since 1860.

*** Deer Creek,** cor.

See *Door Creek*, cor.

Deer Creek

Origin unknown. This is a recent name, found on maps but not used by the local inhabitants. It seems to have been used first on Hiway maps; earlier names were *Middle Branch* (Gree. 1855, Butt. 1880, and all platbooks from 1890 forward), and *North Branch* (Foote 1890) of Sugar R. The stream has two branches rising W in SPRINGDALE, meeting in NW sec. 33, passing thru Mt. Vernon, and joining the West Branch of Sugar R., on sec. 13, PRIMROSE.

Deerfield, town

For *Deerfield PO.* It was est. Mar. 17, 1849, by separation from COTTAGE GROVE; it is T7N,R12E.

Deerfield, vill.

For the PO, around which the original village grew, at the crossroads in SE sec. 9, DEERFIELD (formerly, *Hyer's*). This was never platted. When the C and NW railroad came thru the town (1882) and established a station in SE sec. 21, a new village was platted, including parts of secs. 21 and 28, and the name was transferred here. It was incorporated 1891. The original settlement is called Old Deerfield.

Deerfield Corners

=*Old Deerfield.* Used locally.

Deerfield PO

Said to refer to the many deer found in the fields when the region was first settled (so Park, 432, 436, and others after). But since the name was given by David R. Hyer, the first postmaster, a native of Vermont, it is also very likely (in part at least) reminiscent. (There were previous Deerfields in N. H., Mass., Va., Ohio, Ind., and Mich.) (MCT, 1.3.61, p. 6)

Est. Sept. 12, 1846, and kept at the crossroads in S sec. 9, DEERFIELD. Around it "Old Deerfield" grew up. When the present village of Deerfield was founded on secs. 21 and 28, the PO was moved there.

DeForest, vill. [dɪ'fɑrɪst]

For Isaac N. DeForest, who had acquired the land in 1856. When the CMSP railroad came here, the plat was made (1874) and the name conferred on the new village, covering adjoining parts of secs. 17 and 18, WINDSOR.

DeForest PO

For the village. The name was changed from *North Windsor PO* Dec. 3, 1872.

*** DeForest Station**

=The village of *DeForest,* with specific reference to the station of the CMSP railroad, est. here 1871. An early alternative name for the village.

Deneen School [dɛ'nin]

For the Deneen family (see *Deneen Settlement*). In SE sec. 29, VERMONT.

*** Deneen Settlement, The**

For the numerous Deneen family, whose members settled along the present highway F in SW VERMONT, from 1854 (when P. M. Deneen came from Ireland) forward.

*** DeTour River**

Origin unknown; perhaps from a Frenchman's name, or the French word meaning "deviation."

Placed on Mitchell's map of 1835 on the E. Branch of Mounds Cr., either for that branch alone or for the whole creek. This places it in Dane Co., but probably erroneously; it seems to belong to some creek in Iowa Co., probably the present Mill Cr. (see maps of Taylor, I&W, 1835), or else the W. Branch of Mounds Cr. (Farmer, 1830).

Devil's Chimney, The

Subjectively descriptive: a sandstone pillar about 50 ft. high and standing apart from the rocks about it; SE sec. 11, PRIMROSE. The name was probably in use before 1850. One account identifies nearby formations as "the Devil's Washbasin," "the Devil's Bootjack," etc. (Prim. 99), but these probably have had little if any currency.

Devil's Pocket, The

Subjectively descriptive: a very sharp ravine where two branch ravines meet, in S sec. 9, DANE. (Steele)

Devil's Washbowl, The

Former name for the *Devil's Pocket.* (Steele)

Devine School [dɪˈvaɪn]

For the Devine brothers (Joseph, Manus, DeWitt, Bennett, and John) of N. Y., who settled from 1845 on, mostly in sec. 23, OREGON. The school is in W sec. 24. Now closed.

Diamond Valley

For its shape: another name for Vermont Valley, or that part of it in W secs. 12 and 13. The name was current about 1890; at that time it was given to the Diamond Cheese Factory, which still preserves it. (Pauls.)

Dixon School

For Charles Dixon, English settler, who owned adjoining land from 1870 till before 1890. The school is in SE sec. 18, PRIMROSE. Also called the *LaFollette School.*

Dodge School

For G. Dodge and his descendants, who have owned the adjoining land from before 1861. The school is in SE sec. 13, ROXBURY.

*** Dodgeville Road, The**

An early alternate name for the *Mineral Point Road.* (Park 441)

Dogtown, cor.

From the numerous dogs which a local dweller is said to have kept, and which were an annoyance to passing horses. (However, this name is found elsewhere, derisively used to suggest insignificance—compare a "one-horse town.") A crossroads in the E part of secs. 9 and 16, FITCHBURG.

Dogtown School

For *Dogtown.* It used to be about half a mile N of the crossroads, but is now at it. The name has been officially changed to *Fish Hatchery School*, but is still used locally.

Donald Rock, The

For John S. Donald, who, some time in the 1890's, became owner of the land (SW sec. 28, SPRINGDALE) in which this rock-formation stands. This is the most current name; other names are *McCord Rocks, Picture Rocks.*

Door Creek

Origin uncertain; see *Skeneda Cr.* Possibly the "door" refers to a narrow passage which Little Door Creek makes between high bluffs in the N part of sec. 32, COTTAGE GROVE, where the present crossroads also is (Thor.). The first appearance of the name is on Lapham's map of 1849, tho it had been used before (the PO named for it was est. 1847); significantly, perhaps, this map has the name on the present Little Door Creek. It rises in BLOOMING GROVE (NE) and COTTAGE GROVE (NW), flows S thru the latter into PLEASANT SPRINGS and DUNN, and into the N side of L. Kegonsa.

Door Creek, cor.

For the PO: a small crossroad settlement, originally in E sec. 33, COTTAGE GROVE; this was at first known as *Buckeye*, but the PO name gradually superseded this. Some time between 1873 and 1890 the roads were relocated along the section lines, moving the settlement E to its present place between secs. 33 and 34. On some early maps (Lapham 1849, 1852) it is shown erroneously as "Deer Creek."

*** Door Creek PO**

For the nearby stream, *Door Cr.* Est. Feb. 11, 1847, disct. Sept. 16, 1895; re-est. Mar. 4, 1898, disct. Oct. 15, 1902. Kept near *Door Creek* corner, and later moved E with it.

Door Creek School

For the creek, which flows by just to the W. Between secs. 5 and 6, PLEASANT SPRINGS.

Dorn's Creek [dɔrnz]

For the Dorn family, across whose lands it flows: a local name for *Spring Creek* in WESTPORT. (Corc.)

Dorn's Spring

For P. J. Dorn and his descendants who, since before 1873, have owned the land on which it rises, NW sec. 30, WESTPORT. (Corc.)

Drammen Valley ['dramen]

For Drammen, Norway, from which many of the early settlers came: a valley in S sec. 30, PERRY. Named only on Lig. map, 1861; the name is preserved in that of a cheese factory.

Drainage Ditch, The

= *Big Ditch.*

Dreher School ['dreər]

For Frank Dreher, who bought the adjoining land (sec. 10) before 1899; in SW sec. 3, OREGON. Now closed.

*** Duck Creek**

Erroneously placed on *Black Earth Cr.* on Chan. map (1829).

*** Duck Lake**

An early name for *Lake Wingra.*

Dunkirk, town

For the village. Est. Feb. 2, 1846; T5N,R11E.

Dunkirk, vill.

Probably for Dunkirk, Chautauqua Co., N. Y.: a "paper town," platted Dec. 5, 1836, by Jonathan Parsons, Jr., including the E half of the NW quarter of sec. 28, in what later became DUNKIRK. See *Wood's Addition* and *Dunkirk Falls*, which adjoined it. Settlement in this locality did not begin till 1843. The village is now virtually defunct, and the name falling into disuse.

*** Dunkirk Falls,** vill.

After *Dunkirk* village: a "paper town," platted Dec. 19, 1836, including the S half of sec. 21, the W half of the NW quarter of sec. 28, and the NE quarter of sec. 28, in what became DUNKIRK. It thus enclosed Dunkirk on 3 sides, the two together covering a whole section of land. Most of this ambitious scheme lapsed, tho the plat was not formally vacated.

*** Dunkirk PO**

For the village. Est. Mar. 8, 1847; disct. Mar. 26, 1872. It was kept in the settlement, W of the road, in the SW corner of sec. 21, DUNKIRK.

*** Dunkirk Precinct**

Park (545) mentions an election precinct as having been est. by this name between 1839 and 1846. Since none such is mentioned in Dane, however, Park was probably referring to *Clinton Precinct.*

*** Dunkirk School**

For *Dunkirk* village, in which it was, in SW sec. 21, DUNKIRK. About 1930 this name was changed to *Riverview School.*

Dunlap Creek

For Adam Dunlap, of N. Y., and his family, early settlers (he came in 1848) and owners of adjoining land till the present (secs. 36 and 1, MAZOMANIE). The creek rises in the SW corner of ROXBURY, flows E into MAZOMANIE, then N into the Wisconsin R. Formerly called *Boiling Cr.* See *Big Ditch.*

Dunlap Hollow School

For its location in *Dunlap Hollow;* in SW sec. 31, ROXBURY.

Dunlap('s) Hollow

As the valley of *Dunlap Creek,* where the Dunlap family was among the first to settle.

Dunn, town

By an error for Door, or possibly Dover; when the name was submitted, it was so poorly written that the engrossing clerk of the assembly mistook it and recorded "Dunn," which remained. "Door" is given as the intended name by H. & W. Atlas, 1873, the earliest source, and also by Durrie, 1874; "Dover" is first given instead of Door by Butt., 1880, and later writers have followed one or the other without explanation. Door, besides being the earlier mentioned, seems the more probable because Door Creek, on which was one of the earliest settlements, enters L. Kegonsa in this town. On the other hand, Dover was a common name farther E and S (there were already 14 Dovers by 1842) and may have been brought by early settlers. (Another, but doubtful, explanation has it that it was named for Capt. Chas. Dunn, a pioneer Wisconsin attorney, who died here during the Black Hawk War; see MCT 10.16.38.) The town was est. Mar. 11, 1848, and includes T6N,R10E.

*** Dunn PO**

For the town. Est. Feb. 15, 1849, with John Paine as postmaster; name changed to *Ancient PO*, Jan. 14, 1850.

Dunphy's Corners ['dʌmfɪz]

For J. Dunphy and his family, who settled there in the 1870's: a crossroads in S secs. 11 and 12, SUN PRAIRIE.

Dunrovan [ˌdʌn'rovən]

A pun on "done rovin'," which the present owner, Ralph Mathasen, says he is: the new name of the former *Montjoy.*

Duscheck's Lake ['duʃɛks]

For Mr. and Mrs. Ed. Duscheck, who owned land on its NW side from before 1911 till after 1931: another name for *Brazee's L.*

*** Dwight School**

For Theodore and Edward Dwight and their families, who owned land in secs. 26 and 35, OREGON, from about 1856 till after 1904. The school was on the town road between these sections; now discontinued and removed.

Dyreson's Hill ['diɚsənz]

For Brickt Dyreson and family, owners of the land since the 1890's: a hill in SE sec. 22, DUNN.

Eagle Heights

Subjectively descriptive: a high point above the S shore of L. Mendota, in the NE corner of sec. 17, MADISON. So named in Foote, 1890, and still so known.

*** Eagle Point** (MADISON)

Origin unknown. Recorded only by J. G. Knapp in 1872, as an early name for *Maple Bluff*. (WHS *Coll.* 6.373)

Eagle Point (WINDSOR)

Origin unknown; perhaps reminiscent. It was well known in settlement days because there was an inn or tavern there, the next stagecoach stop N of Token Creek on the Madison-Portage run. It was probably named by William Lawrence of Vermont who took up the land (NE sec. 5, close to the line of sec. 4) in 1841, and kept the inn. N. N. Pike succeeded him (1844–5), and later Jesse W. Helden, under whom it became the Eagle Point Farm, which it still is. (Park 250, 253)

East Blue Mound

See *Blue Mounds, The*.

East Blue Mounds

For the eastern part of BLUE MOUNDS; e.g., in the name *East Blue Mounds Church.*

East Blue Mounds Branch (of Pecatonica R.)

= *Blue Mounds Branch;* a former name, now unofficial, but still in use.

East Blue Mounds Church

For its location in BLUE MOUNDS: a Norwegian Lutheran church in S sec. 23; built 1868 (Barton 1085).

East Blue Mound(s) Creek
=*East Branch (of Mound Creek)*—a former name (Butt., Keyes, etc.).

*** East Blue Mound(s) PO**
An unofficial name for *Blue Mound PO*, applied locally to distinguish it from *West Blue Mound PO*, while the separate PO's were in existence (i.e. until *Blue Mounds PO* was est.). (So H&W)

East Branch (of Mounds Cr.)
Descriptive. It rises in the N part of BLUE MOUNDS, flows north thru VERMONT and BLACK EARTH, and into Iowa Co., where it joins with the West Branch (USGS). Formerly called *Waurooshic*, and *East Blue Mound(s) Creek;* now also *Blue Mound(s) Creek*.

East Branch (of Pecatonica R.)
=*Blue Mounds Branch;* a former name, from 1831 (Mitch.) forward, now unofficial, but still in use.

East Bristol, cor.
For its location, on parts of secs. 2 and 12, BRISTOL. The name did not achieve official status till the establishment of the PO (1879), but the settlement existed much earlier—from at least the 1850's.

*** East Bristol PO**
For the village (at which it was presumably kept, tho Butt. map, 1880, shows it at York Center). Est. June 18, 1879; disct. Aug. 15, 1907.

*** Eastern Branch (of Sugar R.)**
=The present official source branch of Sugar R., evidently not yet thought of as such when this term was used to distinguish it from the *West Branch*. Used only in the Treaty of 1829.

East Koshkonong [ˈkɑʃkəˌnɑŋ]
The E part of *Koshkonong Prairie* or the settlement on it—e.g., in the name *East Koshkonong Church*.

East Koshkonong Church
See preceding entry: a Norwegian Lutheran church in SW sec. 26, CHRISTIANA. The first, of logs, was built in 1844; the present stone church replaced it in 1858. (Barton 1092)

Eastland Park, subd.
For its position within the city of Madison; platted 1930.

Eastlawn Park, subd.

For its position within the city of Madison; also probably to correspond with the earlier West Lawn and North Lawn; platted 1929.

*** East Madison**

From its location with respect to *Madison.* A "paper town" shown on the NE corner of "Third Lake" (L. Monona), in what is now BLOOMING GROVE. (Judson 1836). No plat was recorded.

East Madison

Local name for the eastern part of the city of *Madison;* used as early as 1880 (Butt. 1080).

East Middleton, cor.

For the PO: between secs. 22 and 27, MIDDLETON. Following the names of its various PO's (which see), it has been called Middleton, Middleton Junction, and East Middleton. The latter 2 names are still in use.

*** East Middleton PO**

By contrast with *West Middleton PO,* 2½ miles to the W. The name was changed from *Junction PO* to this, Jan. 17, 1872; disct. Dec. 22, 1893.

*** East Rutland PO**

For its position in RUTLAND. Est. Apr. 25, 1848; disct. July 5, 1848; kept by James H. Cummings.

East Side Addition, subd.

For its location within the village of Belleville; platted 1919.

Ebenezer, subd. [ˌɛbɛnˈizɚ]

For their grand uncle, Ebenezer Brigham, first white settler of Dane Co.; platted 1932, by Charles I. Brigham, Mrs. Katherine B. Fox, and Mrs. Mary R. Brigham, in NW sec. 5, BLUE MOUNDS. Mrs. Fox decided on the name. (Brig.)

Eby's Hill [ˈibɪz]

For R. C. Eby, owner of the land on its S side from before 1931: a sharp hill over which passes the section-line road, between secs. 10 and 15, FITCHBURG.

Edgerton Beach Park, subd.

For the nearby village of Edgerton (Rock Co.); on L. Koshkonong, platted 1926, in SE sec. 25, ALBION.

Edgewater, subd.

For its position along L. Monona; platted 1912, in NW sec. 25, MADISON.

Edgewood Park, subd.

For adjoining "Edgewood," the Dominican property, itself descriptive of its being wooded, and on the edge of L. Wingra; platted 1907, in NE sec. 28, MADISON; since absorbed into the city of Madison.

Edwards Park, subd.

For E. N. Edwards, owner of the land; platted 1899, in adjoining S parts of secs. 33 and 34, BLOOMING GROVE.

Eken Park, subd. [ˈikɛ˖n]

For the Eken brothers (Thos. I. and Ole), owners of the land; platted 1924, in SE sec. 31 and SW sec. 32, BURKE; since absorbed into the city of Madison.

Elder Bowman's Bluff

For Elder (the Rev. Joseph) Bowman and his family, who owned the land from before 1861 till after 1926: a long bluff or range of bluffs towering above county highway Y, in N secs. 4 and 5, DANE. (Steele)

Ella Wheeler Wilcox Creek

Because it passes near the childhood home of Ella Wheeler Wilcox: a creek rising on sec. 35, VIENNA, and flowing SE into the Yahara R. in sec. 31, WINDSOR.

Ella Wheeler Wilcox School

Because it is close to the childhood home of E. W. Wilcox, and is a successor of the school she went to (Barton 1090). On the S edge of sec. 36, VIENNA.

Elmhurst, subd.

Because of the elm trees here; probably also reminiscent of some other place bearing the name. Platted 1926, in SE sec. 17, BLOOMING GROVE.

Elm School

For nearby elm trees: at the road junction in the center between secs. 10 and 11, DANE.

Elmside, subd.

Presumably descriptive, tho the name may have been suggested by the fine home of Dr. J. B. Bowen, "Elm Side," in the SW part of Madison. This subdivision, however, was platted (1891) to the E of the city, in adjoining parts of secs. 5 and 8, BLOOMING

GROVE. It was later absorbed into *Fair Oaks*, and with it into the city of Madison. The name is preserved in that of a boulevard.

Elvers, vill.

For Charles Elver, who, in 1878 bought Saml. Lattimer's grist mill, sec. 21, VERMONT, added to it, and made a considerable local reputation for several years, during which the settlement grew, acquired a PO, etc. The mill burnt before 1890, however, and the settlement has virtually disappeared, tho the name is still used.

Elvers Creek

For the settlement called *Elvers*, tho there is some confusion as to the creek itself. In local usage, the name is always applied to the creek on which the settlement was made, which seems right enough. But beginning with the USGS map (1919–20), the name was placed on the branch next west of this, which joins it in N sec. 17, VERMONT, while the branch on which the settlement is is marked "East Branch" (of Mound Creek). Hiway maps have followed USGS in thus transporting the name "Elvers Cr." 2 miles westward.

*** Elvers Mills**

An early alternate for *Elvers* (which is perhaps an abbreviation of it).

*** Elvers PO**

For *Elvers* village. Est. July 19, 1878, and kept at his store on sec. 21, VERMONT, by John Lohrs, the first postmaster. Disct. Sept. 16, 1899; re-est. Oct. 26, 1899; disct. June 31, 1901.

Emerson Park, subd.

Named by his daughters, who had it platted (1901), for Moses Emerson, a former prominent citizen; within the city of Stoughton; NE sec. 7, DUNKIRK. (Dow)

Emery School ['ɛməɪ, 'ɛmrɪ]

For J. Q. Emery, who bought adjoining land before 1873, and became State Superintendent of Education. The school is in NE sec. 20, ALBION.

*** English Settlement, The**

For the derivation of the settlers: an overflow of the British Temperance Emigration Society's settlement in Mazomanie, in the NW part of BERRY, the SW part of VIENNA and adjoining NW part of WESTPORT. The settlement began in 1846, but did not remain homogeneous much beyond the 70's. (Park 266, etc.)

*** Ensign Hill**

For the owner of the land on which it stood, in sec. 10, BURKE. (So says Park, 426; no other mention of it has been found, and the platbooks show no such owner of land in this section from 1863 forward. Very doubtful.) See *Hanson's Hill.*

*** Eolia PO**

Origin unknown; possibly based on the name Æolus or Eolus, Roman god of the winds? (Eolia, in Pike Co., Missouri, is said to have been so named about 1880 by the surveyor because the day on which he surveyed it was extremely windy. Leech, in *Univ. of Mo. Studies,* Jan. 1934, 74.) Est. July 16, 1850; disct. Mar. 14, 1866. Kept in NE sec. 5, YORK, by Charles D. Bradley, the postmaster.

Erbe School [ɝb]

For Christian Erbe and his family, owners of the land (from before 1873) on which the school is: in SE sec. 9, BLUE MOUNDS.

Erickson Park, subd.

For Edward Erickson, one of the owners of the land, who had it platted 1907; NE sec. 7, DUNKIRK.

*** Eshunikede Lake**

From an Algonquian word (perhaps Potawatomi, less likely Ojibwa) meaning "there is a portage" (Geary). Used only by Taylor (1838) for *Crane L.*

Estes School ['ɛstɪz]

For A. G. Estes, who owned adjoining land from before 1890 till after 1904. The school is in E sec. 11, BLOOMING GROVE.

Esther Beach ['ɛstɚ]

For Esther, the daughter of Charles Askew, one of the owners: a summer resort on L. Monona, never platted; S sec. 19, BLOOMING GROVE. (Woodw., etc.) The name was given some time between 1904 and 1911; it is now remembered widely, but has been superseded by *Hollywood at the Beach.*

Ethelwyn Park, subd. ['ɛθḷwɪn]

For Miss Ethelwyn Anderson, a former local resident; on L. Monona, platted 1900, in S sec. 19 and N sec. 30, BLOOMING GROVE.

Evergreen Park, subd.

Descriptive; within the city of Madison, platted 1903.

* **Everson's Point** ['ivərsənz]

For R. Everson and family, who owned it from before 1873 till after 1904: the piece of land (SW sec. 13, DUNN) jutting into L. Kegonsa, S of the entry of the Yahara R., now the site of *Grand View*.

Fairchild School

For Burke Fairchild and family, among the earliest settlers of the town (1845); he was the first town chairman. The school is in NE sec. 19, ROXBURY.

* **Fair Grounds PO**

For its location near the "Fair Grounds" in SW sec. 25, MADISON. The name was changed to this from *Assembly PO* Oct. 13, 1899; it was disct. Dec. 12, 1899.

Fairhaven, subd.

Subjectively descriptive; perhaps also influenced by the name of a nearby farm; see *Springhaven*. Platted 1911, in adjoining parts of secs. 17, 19, and 20, BLOOMING GROVE.

* **Fair Oaks,** vill. ['fɛrˌoks]

Subjectively descriptive; platted 1901, in adjoining parts of SE sec. 6 and SW sec. 5, BLOOMING GROVE; since absorbed into the city of Madison, tho the name is preserved in that of an avenue.

Fair View, subd.

Subjectively descriptive; an addition to South Madison (and with it absorbed into the city of Madison); platted 1907.

Fairview, subd.

Subjectively descriptive; an addition to the village of *Waunakee;* platted 1903.

* **Fairy Lake**

An early name for *L. Monona.*

* **Farmersville,** town.

Probably for some former town or village of the same name; there were such in Ala., La., Ohio, and N. Y. (Cattaraugus Co.). The last is the most likely source, since many New Yorkers came early to this region. A former town in Dane Co., est. Aug. 2, 1848, and comprising T8N,R6 & 7E. Two days later was added the fractional part of T9N,R6E lying S and E of the Wisconsin River. This addition was transferred, however, to *ROXBURY*,

Jan. 7, 1850, and at the same time it lost T8N,R7E when *BERRY* was formed. Finally, the name was changed to *BLACK EARTH* Feb. 1, 1851, and "Farmersville" became extinct.

Farwell's Point ['farwelz, 'ferwelz]

For former Gov. Leonard J. Farwell, who owned the land, and retired there after 1857: a high "point" along the N shore of L. Mendota in NE sec. 34, WESTPORT. Formerly *Point Mendota.*

*** Father Inama's Creek**

= *Inama's Creek.*

Findlay Park, subd. ['findlı, 'finlı]

For Alexander Findlay, prominent grocer, in Madison from 1863: a replat (1925) of parts of *Quarrytown* and *Lakeland,* in NW sec. 21, MADISON.

*** Finlayson's Hill**

For M. and W. Finlayson, who in the 1870's owned adjoining lands in NW sec. 4, VERMONT, over which present highway F passes. (Linc.)

First Lake

An early name for *L. Kegonsa.*

Fisher Valley

For J. W. Fisher and family, of N. Y., who opened a farm in SE sec. 21, in 1847: a valley in secs. 21 and 22, OREGON.

Fisher Valley Hill or Ridge

For its position with respect to *Fisher Valley:* it curves around E and S of the valley, thru secs. 22, 27, and 28, OREGON.

Fisher Valley School

For its location in *Fisher Valley.* In SE sec. 21, OREGON.

Fish Hatchery Road

For the State Fish Hatchery (est. 1876–7) in NW sec. 10, FITCHBURG, by which it passes. It connects Madison with Oakhall, etc.

Fish Hatchery School

For the State Fish Hatchery, a mile N. Formerly *Dogtown School;* SE sec. 9, FITCHBURG.

*** Fish Lake** (DUNN and PLEASANT SPRINGS)

An early name for *L. Kegonsa.*

*** Fish Lake** (DEERFIELD)

= *Goose Lake;* so only on Lapham's maps, 1849 ff.; probably a mapmaker's error, by confusion with next.

Fish Lake (ROXBURY)

For the numerous fish it once contained: a lake chiefly on sec. 3, also partly on sec. 4, ROXBURY; also a small part of Columbia Co. First on a map in 1859 (Can.); but see *Goose L.* Also formerly called *Clark's L.*

*** Fish Trap Ford**

For an Indian fish trap constructed just above it: an early crossing of Sugar Cr. (in the S part of present sec. 34, MONTROSE) by Indian trails, "1st Lake Trail" (i.e., that to L. Kegonsa) and "4th Lake Trail" (i.e., that to L. Mendota) joining on the E bank of the stream and crossing as one to the W bank. So shown on the map of Lorin Miller, in his notebook made in the survey of 1833.

Fitchburg, town

For the PO and Precinct. The town (T6N,R9E) had been organized as *Greenfield*, Feb. 11, 1847, but the name was changed to Fitchburg Apr. 2, 1853.

Fitchburg, vill.

See *Fitchburg Station.*

Fitchburg Center School

For its proximity to the center of FITCHBURG. Another name for *Gorman School.*

*** Fitchburg Corners**

For the PO: the crossroads in the center of sec. 33, FITCHBURG, later called *Oakhall.*

Fitchburg PO

The name is said to have been suggested by Ebenezer Brigham of Blue Mounds (Park 448); it is therefore probably reminiscent of Fitchburg, Mass., which is in Brigham's native Worcester Co., and not far from Shrewsbury, his birthplace. The PO was est. Aug. 26, 1842, and kept half a mile S of the "Corners" in the center of sec. 33, FITCHBURG (*Oak Hall*). Disct. Jan. 2, 1879. Re-est. and moved to Fitchburg village in SE sec. 30, July 26, 1888.

*** Fitchburg Precinct**

For the PO: an election precinct est. Apr. 2, 1844. The previous name had been *Quivey's Grove Precinct.*

Fitchburg Road

For the settlement (early site of *Fitchburg PO*) which it connected with Madison. During the 1840's it ran from this settlement NW to the site of the present Fitchburg village, and thence as at present to Madison (Park 448). This latter part still bears the name.

*** Fitchburg Station**

For the town: a station on the I. C. railroad, in SE sec. 30, FITCHBURG. The PO was moved here when re-est., 1888, and a small settlement grew up—also called simply "Fitchburg." The station has been discontinued.

Five Corners

A current alternate for *Five Points*. (F. Stew.)

Five Points

Descriptive: the point at which 5 roads meet, at the middle of the line between secs. 3 and 10, VERONA. (USGS 1906–7)

Flint School

For the Flint family, whose members (James, F., etc.) settled before 1861 in the N part of RUTLAND. The school is in S sec. 4.

*** Floral Hill**

Subjectively descriptive: a hill in the north part of secs. 27 and 28, BURKE; in the 1860's the site of a nursery, in the 1890's of a stock farm, to each of which it gave its name.

*** Floyd,** cor.

For *Floyd PO*, which was kept at this place (*Oak Hall*) sec. 33, FITCHBURG. Tho the PO was discontinued in 1902, maps still showed "Floyd" as if it were the name of the settlement—which it never was.

*** Floyd PO**

Chosen by the PO Dept. from a list of several monosyllabic names submitted by Hec Clapp, the prospective postmaster; this was the name of a young farmer of the neighborhood, Floyd Gurnee. As the PO was to be kept at *Oak Hall*, this latter name was first submitted, but it was rejected because there was already an Oak Hill PO in Wisconsin, and confusion between the two was anticipated. As it happened, the name "Floyd" was confused anyhow—with Lloyd PO. Est. Apr. 23, 1893; disct. Sept. 15, 1902. (Barry, Bent., Clapp)

Fond du Lac Road, The [ˈfɑndʊˌlæk]

For its northerly terminus: an early name for the *Columbus Road*.

*** Fordville,** cor.

For the brothers, Robert, Archibald, David, and C. Ford, early influential there: a crossroads settlement in the N part of secs. 1 and 2, SPRINGFIELD; so called from the 1850's; still remembered, tho out of use.

Fordville School

For *Fordville*, at the site of which it is: at the junction of roads between N secs. 1 and 2, SPRINGFIELD. Now closed.

Forest of Fame, The

A park in *Mt. Vernon*, containing trees from several parts of the world honoring Washington, Theodore Roosevelt, Pershing, Lincoln, Robin Hood, William Tell, LaFollette, etc. The Hon. John S. Donald (see *Donald Rock*) was responsible for the idea; the park was begun in 1916.

Forest Park, subd.

Descriptive; also having reference to the earlier plat of Lake Forest, here; platted 1934, in adjoining parts of secs. 27 and 34, MADISON.

Forks, The

Descriptive: the confluence (sec. 16, RUTLAND) of Oregon Branch and Rutland Branch to form *Badfish Creek*.

*** Fort Winnebago Road, The**

Because it connected Madison with Ft. Winnebago (now Portage): an early name for *Token Rd.* (1843: Dane 71)

Forward, vill.

For the PO: a small settlement, mostly in NW sec. 23, PERRY.

*** Forward PO**

Probably from the motto on the seal of the State of Wisconsin. Est. Mar. 4, 1872, and kept at first on the land of Christian Evenson, the postmaster, NW sec. 23, PERRY. After a few years, with a change of postmaster, it was moved into NE sec. 22. It was disct. Aug. 30, 1902.

Forward School

For the village; in NE sec. 22, PERRY.

Four Lakes, (The)

From their number and relationship, forming a chain connected by the Yahara R.: the collective name of lakes *Kegonsa*, *Waubesa*, *Monona*, and *Mendota*, and, by extension, of the region about them. This name is probably a translation, direct or via French, from the Winnebago *Taychopera.* First appears in MS. 1817; in print 1829. The first individual names of these lakes were, respectively, First, Second, Third, and Fourth Lake, most probably given by the surveyors of this region as they worked northward from their base line, the northern boundary of Illinois, in 1833–4. These seem to have been the only names in use by the white men before 1849. For further details, see Cas. 1.

*** Four Lakes River**

=River of the Four Lakes, now the *Yahara R.* This form used by A. M. Mitchell, 1839. (Keyes 128)

Fourth Lake

An early name for *L. Mendota.*

Four-X, (The) (also, **Four-X Corners**) ['foɚ,ɛks]

Because two roads cross, forming an X and four corners: a crossroads in SE sec. 1, WESTPORT. The name (in use about 50 yrs.) was early taken by a saloon here, but whether the saloon or the crossroads had it first is uncertain (Corc., Ellis; Rural map). See *Whiskey Creek.*

Fox Bluff

See *Foxes' Bluff:* a subdivision using this form of the name was platted here (adjoining parts of secs. 4 and 5, MADISON) in 1942.

Foxes' Bluff (also, **Fox's ——, Fox ——**)

For the many foxes found here, according to Park (301), who was first to mention it (1877) and who, in line with his explanation, spells it "Foxes' Bluff"; this explanation is also given in the locality today (Briggs, Corc.), and since there is no record of anyone named Fox ever having lived or owned land here, it may be accepted: a prominent bluff forming a "point" in NW L. Mendota, sec. 5, MADISON.

*** Fox Settlement, The**

For Dr. Wm. H. Fox, his brother George, and their families, who early opened farms (1843), in sec. 35, FITCHBURG.

Fox's Point

The point in L. Mendota formed by *Foxes' Bluff.*

*** Foxville,** vill.

For Abijah Fox, proprietor, who platted it Jan. 30, 1857; in the NW quarter of sec. 3, CROSS PLAINS. This was the third of the villages, parts of which combined to form the present *Cross Plains;* it was known locally as "lower town."

*** Frawley Hill**

For Thomas Frawley, who owned the land from about 1870 till after 1890: a part of the Sand Ridge in NW sec. 16, VERMONT, over which the road climbs. (Linc.)

Frey School [fraɪ]

For A. Frey and his descendants, who have owned the land on which the school is from before 1861, NE sec. 21, ROXBURY.

*** Frog Pond School**

Because there was a large and populous frog pond nearby. The former name of *Oak Lawn School* (DUNKIRK).

Frost Island

For R. D. Frost (see *Frost's Woods*): a small island in the E side of Squaw Bay, W sec. 20, BLOOMING GROVE. It is part of the village of Monona.

Frosts Woods, subd. (also, **Frost's ——**; **Frost ——**)

For R. D. Frost, former owner of the land; platted 1930, in W sec. 20, BLOOMING GROVE. The name was in use long before the plat was made, Frost having settled there in 1850.

Frost Woods Heights, subd.

For the adjoining *Frost's Woods;* platted 1944, in S sec. 20, BLOOMING GROVE.

Fuller's Woods, subd.

For E. M. Fuller, owner of the land (NE sec. 12, MADISON). Platted 1924, and since absorbed into *Maple Bluff* village. The name was probably in use before the platting, and is still current.

Gallagher Park, subd.

For Clyde A. Gallagher, owner and promoter; platted 1925, in SE sec. 32, BURKE, and adjoining NE sec. 5, BLOOMING GROVE.

Gallagher's Garden, subd.

For Clyde A. Gallagher, owner and promoter; platted 1925, in NE sec. 4, BLOOMING GROVE.

Gammon School (sometimes, **Gammons**) ['gæmən]

For John and Warren Gammons and their families, who settled before 1861 on the land where the school is, and remained till the 1890's. The school is in SW sec. 25, MIDDLETON. Also called *Hallock School.*

Ganser Hills ['gænzɚ]

For Henry Ganser, of Prussia, who bought this land soon

after 1873, and whose family have held it since: a prominent stretch of bluff in NW sec. 18, ROXBURY.

Garden Homes, subd.

Subjectively descriptive; platted 1926, in SE sec. 17, MADISON.

Garden Ridge Park, subd.

Subjectively descriptive; platted 1918, in SW sec. 5, BLOOMING GROVE; since absorbed into the city of Madison.

Garfoot School ['gar,fʊt]

For Wm. Garfoot, English settler, and his family, who have owned land here since 1851. The school is in N sec. 17, CROSS PLAINS. Formerly *Braun School.* Now closed.

Gaston Corners ['gæstən]

For Albert Gaston, of Mass., and his family, who have owned land there from at least the 1850's: a crossroads in N sec. 4, COTTAGE GROVE.

Gaston School

For the Gaston family. At *Gaston Corners.*

Gay Spring

For L. W. Gay, who (before 1911) owned the farm on which it rises: another name for *Vilas Spring.*

German Prairie

Because it was chiefly settled by Germans: a prairie in NE WINDSOR and adjoining parts of Columbia Co. Formerly so called in tho "Yankee" settlement in the S part of the town. The part in Dane Co. is now usually known as *North Windsor.* (Ellis)

*** German Settlement, The**

Descriptive; so called in early times by the "Yankees" in other parts of the town: the settlement in W COTTAGE GROVE, where *Hope* later grew up (Cunn.). The Immanuel Lutheran Church was founded there, 1854.

German Valley

For the national origin of the local settlers (in an otherwise Norwegian-settled area): a small valley, chiefly in secs. 21 and 28, BLUE MOUNDS. (USGS 1916–20)

German Valley School

For the valley; the school is in S sec. 21.

Glen Oak Hills, subd.

The latter part is descriptive; the "glen" puns on the geographic term and the name of one of the owners, Glen P. Turner. Platted 1926, in sec. 19, MADISON.

Glenwood, subd.

Subjectively descriptive; "glen," not a native local term, is evidently used for exotic suggestion. Platted 1916, in NW sec. 28, MADISON.

Goff School [gɑf]

For James Goff and his family, who, from before 1873 till after 1904, owned the land on which the school was built. NE sec. 18, DUNKIRK.

Golf Plat, subd.

Because it was alongside of (and owned by) the Maple Bluff Golf Club; platted 1915, in NE sec. 1, MADISON; since absorbed into *Maple Bluff* village.

Golf View, subd.

Because it adjoins the Burr Oak golf course; platted 1929, in parts of secs. 34 and 35, MADISON.

Goltz's Lake [ˈgoltsəz]

For John Goltz and family, who owned the land from before 1873 till after 1931: a small lake in S sec. 6, SPRINGFIELD.

Goose Lake

For the many wild geese frequenting it: a small lake on parts of secs. 1, 2, 11, and 12, DEERFIELD. This was one of the first of the smaller lakes to be noticed by the map-makers, who, however, show it without name from 1835–45 (Terr., etc.). Lapham (1849) first names it *Fish Lake*, probably erroneously. First to name it Goose L. is Lig. (1861). Park (1877) calls it "Goose Pond," which is closer to the truth.

Goose Pond, The

Because it was frequented by wild geese in early days. The pond was shown by map-makers from 1838 (Taylor) onward, tho left unnamed until 1873 (H&W) and again unnamed thereafter. In SE sec. 12 and NE sec. 13, VERONA.

* **Goosh-ke-hawn River** (also, **Goosh ke-Hawn, Gooskehawn**)

A variant of *Cos ca hó e nah R.*; used by Farmer on his maps of 1828? and 1830, and copied by Tanner, 1833.

Gordon School

For T. Gordon and his descendants, who have held the adjoining land since before 1874. The school is in NE sec. 3, VERONA.

Gorham Heights, subd. ['gɔrəm]

For the Gorham family: see *Gorham Spring*. Platted 1911, in SW sec. 28, MADISON. Since absorbed into Nakoma, and with it into the city of Madison.

Gorham Spring

For James W. Gorham, who, about 1860, bought the early Spring Grove Tavern, or Spring Hotel, just across the road. Now within the University Arboretum; SW sec. 28, MADISON.

Gorman School ['gɔrmən]

For Arthur Gorman and his descendants, who have owned adjacent lands since before 1904. At the crossroads, in NE sec. 21, FITCHBURG.

* **Gorstville,** settlement

For Robert Gorst, the moving spirit of a settlement begun in 1843 by the British Temperance Emigration Society, and to which by 1850, nearly 700 emigrants had come. Gorst's home, in what is now BLACK EARTH, was the center of the settlement, which ultimately spread over much of the present BLACK EARTH, MAZOMANIE, and BERRY, and into adjoining Iowa County. The name was used at least as late as 1849, though other names became current before this; see *Pleasant Valley, Farmersville.* (Kittle 19)

Governor's Island

For Governor Leonard J. Farwell, who formerly owned this and adjoining land: originally an island, it became joined to the mainland (N side of L. Mendota; SE sec. 35, WESTPORT) from about the 1860's, forming a peninsula. The name is still used, however.

* **Grace Church**

An inspirational name: a Roman Catholic church in the Irish settlement of N DEERFIELD. It was built before 1873 in SW sec. 9; removed about ten years ago. (Barton, Onst.)

* **Grace PO**

Probably for *Grace Church*. Est. June 28, 1882, and kept by the postmaster, Charles Mayer, in DEERFIELD, probably on his farm on sec. 9. Disct. Dec. 8, 1885.

* **Grand Crossings**

For the crossing or ford of two Indian trails here, later used by the whites; now *Black Bridge* (*Crossing*). The word "grand," in this use, suggests a French origin. It is shown on Judson's map, 1836.

*** Grand Spring** (also, —— **Springs**)

Descriptive: there were two large springs on secs. 23 and 26, MONTROSE, one of which ("Grand Spring," now *Remy Spring*) was particularly known to travelers between the SW Wisconsin lead mines and Madison, since it was close beside the road.

*** Grand Spring PO**

For the nearby *Grand Spring* on the land of its first postmaster, George McFadden. Est. June 4, 1846; name changed to *Paoli PO*, Nov. 27, 1858.

*** Grand Springs Precinct**

For the Grand Spring settlement, in which it centered: an election precinct est. June 29, 1846. (Park 545; not mentioned in Dane)

Grand View, subd. (MADISON)

Subjectively descriptive; an addition to *South Madison* (and with it absorbed into the city of Madison); platted 1889.

Grand View, subd. (DUNN)

Subjectively descriptive; on L. Kegonsa, S of the entry of the Yahara R. to the lake, in SW sec. 13, DUNN. No plat has been recorded, but it is shown on platbooks from 1911 forward.

*** Grass Lake** (DUNN)

For the grass growing in it: a shallow pond in sec. 30, DUNN, now nearly dried up; shown on platbooks, 1861 as a lake, 1890 as a swamp tho still called "Grass Lake"; thereafter it is omitted.

Grass Lake (DUNKIRK)

For the grass growing in it: a shallow lake on secs. 18 and 19, DUNKIRK. Shown without name on many early maps (Terr., 1835, forward); first with the name on H&W, 1873.

Gravel Store School

Because it is at the crossroads (sec. 23, ALBION) where there used to be a store with gravel on the walls, the first store opened between Albion and Milwaukee (Bab.).

Graveyard Hill

For the old cemetery there, now little used: another name for *Kohlman's Hill* (Stew.).

*** Great Cave of Dane County, The**

= *Richardson Cave*. This name was used in print only by Pedigo, probably derisively, since his description of the cave is derisive. It may even be his invention, since the name is now unknown in the locality.

*** Great Lake**

An early name for *L. Mendota.*

Greenbush, subd. (also, **Green Bush**; now humorously, "The Bush")

Descriptive? Platted 1854, in adjoining W parts of secs. 23 and 26, MADISON. Since absorbed into the city of Madison.

Greene School

For the early Greene family; see *Greene's Creek.* Formerly called *Colladay School;* itself superseded by *Lakeside School.*

Greene's Creek

For Asher G. Greene (who settled here before 1873) and his descendants, across whose land it flows: a small stream rising in sec. 28 and flowing into L. Kegonsa. (Coll.) Unnamed on maps, and by some local dwellers.

*** Greenfield,** town

Probably for some settlement farther E, of which there were already at least 10 before this: a former town (T6N,R9E) organized Feb. 11, 1847. Its name caused confusion with another Greenfield (Milwaukee Co.), so it was changed to Fitchburg, Apr. 2, 1853. (Park 456)

Greenridge Park, subd.

Descriptive; on L. Waubesa; platted 1933, in adjoining parts of secs. 8 and 9, DUNN.

Griffith's Beach, subd. ['grɪfɪθs, 'grɪfəθs]

For John Griffith and his descendants, owners of the land since before 1873; platted 1924, in SW sec. 20, BLOOMING GROVE.

*** Griffith School**

For W. H. Griffith, who settled in nearby sec. 6 before 1861: a former name for *Schaller School.*

*** Grit PO**

For the perseverance or "grit" which it took to get it established, despite various difficulties. So named by the postmaster, Ludwig Sutter, Jr., and kept on his land in SE sec. 35, BLUE MOUNDS (Goebel, Sutter). Est. July 20, 1894; disct. Mar. 20, 1896.

Groveland, subd.

Descriptive; perhaps also referring to BLOOMING GROVE, in which it was platted (1916) in adjoining parts of secs. 6 and 7. Since absorbed into the city of Madison.

*** Halfway Creek**

See *Halfway Prairie Cr.*

*** Half-way House, The**

From its location approximately between Madison and Aztalan: an early tavern, built 1839, by Philip J. Kearney on sec. 18, DEERFIELD: the first home built in the town. (Park 432–4)

Halfway Prairie (also, **Half-Way** ——)

Probably for *Halfway Creek*, which runs thru it. First settlement was made about 1844 by Moses Barnes, on sec. 4. It extends E and W thru the N part of BERRY. The name was applied chiefly to the upper part, earlier; it has now been shifted to refer particularly to the lower part, running into MAZOMANIE (Dahmen).

Halfway Prairie Creek

This, the present form of the name, suggests that the creek was named for *Halfway Prairie*, thru which it flows; the reverse, however, was probably the fact. The earliest form was "Halfway Creek," shown on maps from 1849 (Lapham) forward, and probably in use for some years before that by travelers, for whom it marked about the mid-point between Cross Plains and Sauk City. The order of naming seems to have been, then, first the creek, named "Halfway Creek," then from that "Halfway Prairie," and finally (when the prairie became more important in the minds of agricultural settlers?) the creek was renamed to include the prairie. "Half Way Prairie Creek" appears in 1873 (H&W). The creek has its main source in Indian Lake (USGS), and flows W thru BERRY and MAZOMANIE to join Black Earth Cr.

*** Halfway Prairie Ridge**

For its location north of Halfway Prairie: a ridge running E and W (BERRY and MAZOMANIE).

*** Half-way Tree, The**

For its position between Oregon and Madison: a large oak, formerly used as a landmark. It was in the road (now US highway 14) in about the middle of sec. 12, FITCHBURG, and had to be destroyed when the road was paved. (Brown 7)

Hallock School [ˈhælˌɑk, ˈhælək]

For Youngs Hallock and his descendants, who have owned land close to the school since before 1861. Another name for *Gammons School.*

Halunkenburg (also, **Halunkaburg**), cor. [ha'lʊŋken,burg, hə-'lʊŋkə,bɝg]

From German, meaning "Louts'-town": a nickname for Springfield Corners and its neighborhood, in which lived a group of rough and spendthrift farmers who did much drinking and fighting, both in their own locality and in nearby villages, until it came to be said of any disorderly person, "He must be from Halunkenburg!" The name was used in the German settlements of ROXBURY, BERRY, DANE, etc., but was also known to local non-Germans. Less used today than formerly.

Halvorson's Cross Road ['hælvɚsənz, 'hɒlvɚsn̩z]

For Gregor Halvorson, Norwegian settler, and his descendants, who have held the adjoining land in SE sec. 8 since 1847: the crossroads between secs. 8, 9, 16, and 17, PLEASANT SPRINGS.

* **Hamlin PO** ['hæm,lɪn]

Probably reminiscent; there were Hamlin PO's already in 6 states. The postmaster, Fred J. Wiedenbeck, and officers of the York Center Creamery met and put together a list of possible names, from which, when it was submitted to the PO Dept., "Hamlin" was chosen (Wied.). Est. July 27, 1897, and kept at *York Center;* disct. Feb. 28, 1900.

Hammersley Heights, subd. ['hæmɚzlɪ]

For E. C. Hammersley, owner of the land; platted 1907, in NE sec. 35, MADISON.

Hanan's Pond ['hænənz]

For J. C. Hanan, who settled on adjoining land before 1861: a pond in SW sec. 18, RUTLAND, and SE sec. 13, OREGON. (Newton)

Hanan Spring

For Henry S. Hanan, of N. Y., and his family, who settled by it in 1850: a spring in NE sec. 17, RUTLAND.

* **Hanchettsville,** vill. (also, **Hanchettville**, **Hanchett Ville**) ['hæntʃəts,vɪl]

For *Hanchettsville PO* and its postmaster, who also platted this village, 1849. It developed out of *Bird's Ruins,* and later became part of *Howard City* and *Marshall.*

* **Hanchettsville PO**

For Asahel M. Hanchett, its first postmaster. Est. Mar. 10, 1848; changed to *Marshall PO* Feb. 19, 1862.

Hanerville, cor. [ˈhænɚvɪl]

For the PO: a small settlement about the crossroad between secs. 23 and 26, DUNKIRK. The school there still preserves the name, tho the settlement is gone. (In the middle of the crossroads still stands the large oak tree used as a section marker by the surveyors in 1833.)

*** Hanerville PO**

For James Haner, of N. Y., on whose land (N part of sec. 26, DUNKIRK) it was est., Oct. 27, 1865. Disct. Aug. 15, 1900.

Hanerville School

For *Hanerville*. At the crossroads between secs. 23 and 26, DUNKIRK.

*** Haney Road, The**

For Berry Haney, who kept a ferry across the Wisconsin R., 1839–40: an early local name for the *Sauk Road*, which at the time ran from Madison to Haney's ferry. (Dane 66)

*** Haney's** [ˈheˌniz, ˈhenɪz]

= *The Haney Stand* (only Taylor, 1838).

*** Haney's Creek**

For Berry Haney, who settled on it: an early name for *Black Earth Creek* (at least, in the region of Cross Plains). Mentioned only in Durrie (50) and Park (24), in recounting an expedition of 1837. See next.

*** Haney Stand, The**

For Berry Haney, proprietor: a tavern and stopping-place for travelers, by Black Earth Cr., at the foot of the bluff in SE sec. 3, CROSS PLAINS. This was one of the earliest settlements in the region, at a strategic spot along the military road (see *Cross Plains*); Haney came in 1836 and stayed till late in 1838, when he went to Sauk City. (On his return in 1840, he built a new house in sec. 10, which still stands). (Butt., Dahmen)

*** Hanna School** [ˈhænə]

For J. G. Hanna and his family, who owned the land from before 1873 till after 1899: former name for *Primrose Center School.*

Hanover, subd. [ˈhænˌovɚ]

For Hanover, Germany, from which had come Behrend Veerhusen, one of the owners of the land; platted 1856, including the S half of sec. 25, WESTPORT. Approximately the E half of this plat was later vacated.

Hanson's Hill ['hænsn̩z]

For Peter Hanson and his family, owners of the land since before 1890: a low hill in S sec. 14, BURKE.

Happy Hour School

Subjectively descriptive. Another name for *Belda School.*

Happy Valley School

Subjectively descriptive. At the crossroads between secs. 20 and 29, BRISTOL.

Harker School

For Charles Harker and his descendants, who, since 1854, have owned the land on which the school was built. The school is in E sec. 11, PRIMROSE.

*** Harvey PO**

For Gov. Louis P. Harvey, who had just been elected. The PO was est. July 29, 1862, with Charles A. Martin as postmaster, and was probably kept on his land in NE sec. 24, DANE. The name was changed to *Dane Station PO*, Nov. 1871, and the PO was probably moved to this new settlement.

Harvey School

For *Harvey PO*, formerly nearby. The school is in NE sec. 19, VIENNA. Now closed.

Haseltine Marsh ['hezl̩ˌtɑın]

For Curtis W. Haseltine, former owner: an extensive marsh which formerly covered the greater part of NW MAZOMANIE. Haseltine disposed of it in the 1890's, and the name is now known only to the older people. The greater part is also now under cultivation. (Laws, Butz)

*** Helland Spring** ['hɛlˌand]

For O. M. Helland, owner of the land from before 1873 till after 1904; an alternate name for the *Ruste Spring;* the latter has been more widely used from the first, however.

Helland School

For the Helland family (see *Helland Spring*): in W sec. 12, VERMONT. Now closed.

Helms' Hill [hɛlmz, 'hɛləmz]

For Henry Helms, of N. Y., and his family, first settlers on it (1845): a prominent hill, mostly in sec. 21, FITCHBURG; also called *Mount Pleasant* and *Whalen's Hill.*

*** Henderson PO**

For Joseph R. Henderson, the first postmaster. Est. May 9,

1893, and kept at first at his house, in SW sec. 14, SPRINGDALE; later moved to SW sec. 23; disct. Jan. 31, 1900.

Henry Spring

For Joseph Henry (who some time between 1873 and 1890 bought the former James Berg farm) and his family, who still own the land: a large spring, rising in SE sec. 11, MONTROSE, and flowing W into Sugar R. Formerly *Berg Spring.*

Hessian Hollow ['hɛʃən 'halo]

= *German Valley* (Brig.); an alternative used occasionally by the native Americans, not in its specific sense, but with Colonial overtones, suggesting suspicion of foreigners.

Hickory Hill School

For its location: it is on Springfield Hill, where there are many hickory trees; SE sec. 36, ROXBURY. The hill itself is not called "Hickory Hill," however.

Hickory Hills Estate, subd.

Descriptive: on L. Koshkonong, platted 1936, in SE sec. 25, ALBION.

Hiddesen's Island ['hɪdəsn̩z]

For Mr. Hiddesen, owner since about 1930: a more recent alternative name for *Mosquito Is.* (Derl.)

Hiem's Woods, subd. ['haɪmz]

For Ferdinand J. and Lena Hiem, owners of the land; platted 1937, in SE sec. 12, MIDDLETON. The name was in local use before the platting.

Hiestand School ['histənd]

For John R. Hiestand, of Ohio (who came here 1864) and his family, owners of the adjacent land till after 1911. The school is in E sec. 4, BLOOMING GROVE.

*** High Bank, The**

Descriptive: a low hill of glacial deposit in NE sec. 12, MADISON, between the bend in Sherman Ave. and L. Mendota; leveled about 1890 and used to fill *Thornton's Marsh.* (Woodw.) A local name.

High Hill, The

Descriptive: a hill whose summit is in SE sec. 31, WESTPORT; it is the highest point around L. Mendota. So called locally for at least 30 yrs. (Briggs)

Highland Park, subd.

Descriptive of its location; platted 1906, in NE sec. 21, MADISON; since absorbed into the city of Madison.

Highlands, The, subd.

Descriptive and probably reminiscent; platted 1911, in SE sec. 13, MIDDLETON, and SW sec. 18, MADISON.

High Lawn, subd.

Descriptive; an addition to the village of Brooklyn; platted 1923, in SW sec. 31, RUTLAND.

* **High Mountain**

Descriptive: an early designation for the *Blue Mounds.*

High School Addition, subd.

For its proximity to the High School; platted 1916, within the village of Deerfield.

Highwood Estates, subd.

Partly descriptive, but "estates" is probably intended to lend distinction; in wooded land above L. Koshkonong, platted about 1927 in NE sec. 36, ALBION.

Hildreth School ['hɪldrɛθ]

For S. Hildreth and his family, who owned the land on which it is from before 1861 till about 1900. In SW sec. 13, RUTLAND. Now closed.

Hill, The

= *Bascom Hill,* in University of Wisconsin student parlance.

Hill Crest Park, subd.

Descriptive; platted 1941, in NE sec. 5, SUN PRAIRIE.

Hillcrest School

Descriptive of its location. In NE sec. 35, RUTLAND. Formerly *Hyland School.*

Hillington, subd.

For its suggestion of England. The name was chosen with others (Rugby Row, Eton Ridge, etc.) from a Baedeker guide book for the streets of the plat (1917) by Alfred T. Rogers, Secy. of the Madison Realty Co. (Brown); in SE sec. 21, MADISON; since absorbed into the city of Madison.

Hillington Green, subd.

For the street and park named Hillington Green in nearby *Hillington;* platted 1921, in SE sec. 21 and NE sec. 28, MADISON; later absorbed into the city of Madison.

Hillside, subd.

Descriptive; within the city of Madison; platted 1907.

Hillside, cor.

For the Hillside Creamery, built beside a hill: a crossroad settlement between secs. 35, CHRISTIANA, and 2, ALBION; a store has been there from before 1911.

Hillside School (CHRISTIANA)

For nearby *Hillside* corner. In N sec. 35, CHRISTIANA.

Hillside School (DUNN)

For its location on *Daly's Hill;* formerly *Daly School.* Now closed.

Hiney's Slough ['hɑɪnɪz 'slu]

For Frank H. Hiney, owner of the land: a backwater in the E channel of the Wisconsin R., E of *Bittersweet Is.*, and W of *Ice* and *Spring Sloughs.* (Derl.)

Hippe's Corner ['hɪpɪz]

For T. A. Hippe, and before him A. L. Van Hippe, who have owned the land NW of it from before 1926: the junction in S sec. 27, ALBION. (Bab.)

Hippe's Hill

See *Hippe's Corner:* a sizable hill on SE sec. 27 and NE sec. 24, ALBION. (Bab.) Formerly *Byron Long's Hill.*

* **Hobart,** town

Origin unknown. Shown only on Greeley's and Morse's maps of 1855, instead of *VERMONT.* Since VERMONT was established and named in that year, it may be that the mapmakers, hearing that the new town was to be called Hobart, recorded that before the actual decision was made. See *SMITHFIELD.*

Hoboken Bay [ˌho'bokən, 'hoˌbokən]

For *Hoboken Beach,* nearby: a small indentation of the shore of L. Monona in SE sec. 19, BLOOMING GROVE.

Hoboken Beach, subd.

Probably for Hoboken Cottage, the home of Robert Price, which was in this vicinity (and ultimately for the city in N. J.); on L. Monona, platted 1904, in SE sec. 19, BLOOMING GROVE, by Fayette Durlin and E. A. Cook, one of whom named it. (Woodw.)

The name "Hoboken Club" appears on a map (Monona) in this location as early as 1900, so the "Hoboken" part of the name was current well before the time of the platting. Because this had been a "tough neighborhood," a folk-etymological explanation has grown up, by which the name is sometimes said to refer to the "hobos" who were there.

Hoboken Point

For *Hoboken Beach:* a small point of land along the shore of L. Monona, SE sec. 19, BLOOMING GROVE. Shown on Rural map, 1910.

Hoepker's Corners ['hɛpkɚz]

For John Hoepker and his family, who, from 1875 or before have owned the adjacent land: the crossroad where secs. 10, 11, 14, and 15 meet, BURKE.

Hoffman Corners

For the Hoffman family, whose members have owned land on two sides of it since at least 1873: the crossroads where secs. 35 and 36, COTTAGE GROVE, and 1 and 2, PLEASANT SPRINGS, meet.

*** Hog's Back, The**

Descriptive: a "hogback" or ridge over which an early road (in use before 1832) from Brigham's mines led to Arena, Iowa Co. If this road, as seems probable, ran where the present County Trunk F does, the ridge in question must be that beginning on the N side of the E Blue Mound and running N thru secs. 32 and 29, VERMONT. Accounts do not identify it clearly (Park 243; Butt. 933; Barton 1037; Smith I).

Hog's Back (Hill), The

Descriptive: a "hogback" or ridge about a mile and a half long, mostly in secs. 25, OREGON, and 30, RUTLAND. The old *Lead Road* ran over part of this. (Newton, Anth.)

Hollywood at the Beach

Presumably for the California city: a resort at the location of the former *Esther Beach* (not to be confused with the Hollywood dance hall near Black Bridge).

Holum's Creek ['holəmz]

For Ole Holum, early Norwegian settler, who from before 1890 owned land in N sec. 17, WINDSOR, thru which the creek runs: a local name for that part of the *Yahara R.* near DeForest. (Linde)

Homestead Highlands, subd.

Descriptive; there are high lands, and there were homesteads here; within the village of Monona; platted 1942.

Hook Lake

From its shape, the S end narrowing and curving eastward: a shallow lake in secs. 28, 29, 32, and 33, DUNN. Shown without name on many early maps, from Terr. (1835) forward. First named on Lapham's map, 1849, but erroneously, "Island L.," which is the name of a smaller lake not 2 miles away. (The error is understandable, since Hook L. does contain *Timber Island.*) First correctly named by Lig., 1861. USGS (1904) shows it as swamp; it is not much more than that today. See *L. Peshugo.*

Hook Lake School

For nearby *Hook Lake.* In SE sec. 31, DUNN. Now closed.

Hope, cor.

For the PO: a small settlement in SE sec. 19, COTTAGE GROVE.

*** Hope PO**

Probably a pious name; the PO was in a very religious community, *The German Settlement,* of which the postmaster, Joseph Keuling, was a member. It was est. Mar. 26, 1888; kept in W sec. 19, COTTAGE GROVE; disct. Oct. 31, 1902.

Hope School

For Hope corner: it is about ¾ mile E of the corner, in NE sec. 30.

Horeb's Corners

For *Mount Horeb,* specifically the crossroads in the E part of the village, about which its growth began: a local variant name.

Hornung's Island ['hɔrnəŋz]

For the Hornung family, in whose hands it has been since Anna Hornung acquired it before 1890. It is shown erroneously on some platbooks; according to the tax rolls of MAZOMANIE, it is the island next below Railway Bridge Island, i.e., on secs. 14 and 23, in the N part of the town. (Derl., etc.)

*** Hororah Lake**

Meaning uncertain: a name, found only on Taylor's map (1838) for *Indian Lake.* Tho in form it seems Winnebago, and tho it was these Indians who frequented the lake, the word must be distorted. Possible sources may be "Horah," the fish, or "Hochungarah," the Winnebago Indian.

Horseshoe Bend, The

Descriptive: a bend in the road between adjoining S parts of secs. 26 and 27, VERONA. (Davids.)

*** Howard City**

Said to have been for an official of the CMSP Railroad, which came thru in 1859: this was hardly more than a "paper town" (tho the plat, 1857, included *Hanchettsville*) projected to include parts of secs. 9, 10, 15, an 6, MEDINA. See *Marshall.*

Howard Place, subd.

For Howard Morris, President of the Howard Place Co.; within the city of Madison, platted 1907.

*** Howarth('s) School**

For Henry Howarth, English settler, to whose land it was moved, in NW sec. 15, MAZOMANIE, in 1850; see *School Section Bluff.*

*** Huber School** ['hjubɚ]

For Joseph Huber and his family, who owned adjoining land from before 1890 till after 1899. A former name for *Magelson School.*

Hudson Park, subd.

For J. W. Hudson, former owner of the land; on L. Monona, platted 1902, in adjoining E parts of secs. 6 and 7, BLOOMING GROVE. Since absorbed into the city of Madison.

*** Hulsether School** ['hʌlˌsitɚ]

For L. L. Hulsether, who owned the adjoining land from before 1873 till after 1904. A former name for *Sunnyside School* (CHRISTIANA), and itself preceded by the name *Vee School.* (Onst.)

Hundred Mile Grove

Because in it was driven the surveyors' stake marking the first 100 miles from Fort Crawford (Prairie du Chien) on the old *Military Road* to Fort Winnebago (Portage), probably 1832. The grove covered adjoining parts of secs. 1, 11, 12, and 13, DANE, and 6, 7, and 8, VIENNA. See *Hundred Mile Tree.*

Another explanation (obviously a later local growth) is that farmers hauling wheat to Milwaukee took this grove as marking 100 miles from that city.

Hundred Mile Grove School

For *Hundred Mile Grove,* which once included its site. At the crossroads, in SW sec. 7, VIENNA. Built 1851.

Hundred Mile Tree

For a large white oak which stood by the old military road from Prairie du Chien to Green Bay, about 100 miles from the former, and therefore an early landmark. One of the trees in *Hundred Mile Grove;* Brown (7) locates it in SW sec. 7, VIENNA, in 1935.

* **Hyer's** cor. ['hɑɪərz]

For David R. Hyer, of Vermont, prominent early settler (1843): the crossroad in SE sec. 9 where the "Hyer Hotel" was, the beginning of the original village of *Deerfield.* "Hyer's" appears on Lapham's map (1848) specifically for the hotel, but the name had likely been extended to the settlement. Deerfield PO was established here in 1846, and this new name doubtless began at once to replace "Hyer's." Finally, Hyer himself moved away in 1854. See *Hyer's Corners.*

* **Hyer's Corners**

For David R. Hyer, its chief figure: a small settlement at the crossroads in the N part of sec. 4, SPRINGFIELD. Hyer came here in 1862, built the Hyer House, and had a PO established with himself as postmaster. The settlement began to be reduced in the 1890's; it has now disappeared.

* **Hyer's Corners PO**

For the settlement. Est. Apr. 1, 1872; disct. Dec. 6, 1887.

Hyland School ['hɑɪlənd]

For F. Hyland and family, who bought the adjoining land some time between 1911 and 1926: a former name for *Hillcrest School.*

Ice Slough, The ['ɑɪs ˌslu]

Because it freezes over entirely (see *Spring Slough*): a back water from Blum's Creek, in SW sec. 7, ROXBURY. So called by sportsmen, from the 19th century. (Derl.)

Idlewild, subd.

Subjectively descriptive and probably reminiscent; platted 1916, in SE sec. 33, BLOOMING GROVE.

Illinois Central Railroad

For its original location. It enters Dane Co. at Belleville, and runs NE to Madison. It was completed in 1888.

* **Inama's Creek** (also, **Father Inama's Creek**) (Locally, in the German-derived community, [i'nɑˌmɑ]; also, [ɪn'ɑmə])

For Father Adelbert Inama, founder of the German Catholic settlement in ROXBURY, who, in 1845, settled on its bank, near which the ruins of his chapel may still be seen, SW sec. 17. Now *Madison Brook* or *Blum's Cr.*

* **Indian Camping Ground**

For its former use by the Winnebago; a part of present sec. 3,

BURKE. White settlement here began 1841, with the coming of G. A. Spaulding; about this spot grew up the village of *Token Creek*.

*** Indian Creek**

Because formerly frequented by the Winnebago: a small creek flowing into L. Mendota from the S; SW sec. 15, MADISON. (Brown)

Indian Heights, subd.

For the mounds and other relics of former Indian habitation; on L. Koshkonong; platted 1931, in SE sec. 36, ALBION.

Indian Hill

Said to be for an "Indian Hill" near Stoughton, DUNKIRK, from which the namer, Thor Kittelson Stolen, had come (Kitls.). (This has not been located, however.) It was named about 1882 when Stolen and others were starting a cheese factory at the foot of the hill. The hill was named, and the cheese factory from it. It is in SE sec. 29, PERRY.

Indian Lake

For the Winnebago Indians, who frequented it for hunting and fishing before and for many years after the coming of the whites. Shown on several early maps from 1837 (Mor.) forward, but unnamed except once: 1838 (Taylor) it appears as *Hororah L.* The name "Indian L." first appears on a map in 1873 (H&W), but was surely in use long before that date. See also *Schumann's L.* and *Stapelman's L.*

*** Indian Spring** (BURKE)

For its having been formerly frequented by the Winnebago Indians, whose main trail from Lakes Koshkonong and Waubesa to Fort Winnebago (now Portage) passed close by: a spring on sec. 27, which forms the head of Starkweather's Creek.

Indian Spring (MADISON)

Because frequented by the Winnebago Indians. A large spring close to the south shore of L. Wingra, now within the University Arboretum; sec. 27. Also called *Big Spring*. (Brown 7.)

*** Indian Village, The**

Descriptive: a village of Winnebago Indians on the NW shore of L. Mendota (W sec. 33, WESTPORT), well known in the region in early days.

Interlake, subd.

For its location between lakes Monona and Waubesa, and

reminiscent of Interlaken, the famous Swiss resort; platted 1911, in adjoining parts of secs. 20 and 29, BLOOMING GROVE, and named by Ray S. Owen.

Interlake Lagoon

For *Interlake,* in which it is: an artificial channel connecting with the Yahara R.

Interlake Park, subd.

For *Interlake,* of which it is virtually an addition; platted 1916, in SW sec. 20, BLOOMING GROVE.

Irish Lane

Because it passes thru a solidly Irish settlement, the part of the section line road (between secs. 14 and 15, and 22 and 23) which stretches from county highway D to the Syene Rd. (Purc., Barry)

Irving Park, subd. [ˈɜviŋ]

Origin unknown. Since it adjoined the Knickerbocker Ice Co., this may possibly have suggested to its promoters Washington Irving, who wrote *Knickerbocker's History of New York.* Platted 1915, in NE sec. 8, BLOOMING GROVE.

Island Lake

For the thickly wooded island near its center: a lake on sec. 3, RUTLAND. First known use of the name is on Lapham's map (1849), but he places it erroneously on the similar, nearby *Hook Lake.* First correct use is on Lig. map (1861). USGS map (1904) shows this lake as merely a swampy depression; it is now almost dried up.

*** Isthmus PO**

Origin unknown; not descriptive of its location, but possibly referring to that of Madison, the nearby capital, noted for this feature; it seems to have been along the first Milwaukee-Madison road, since the postmaster was Benjamin Titus, who owned adjoining land in NE sec. 13, COTTAGE GROVE. Est. May 30, 1849; disct. Jan. 17, 1851.

*** Janesville Road, The Old**

Because it was the stage route from Madison to Janesville. The part leaving Madison was the same as the *Fitchburg* or Stoner's Prairie Rd.; from Fitchburg PO it ran SW thru Rutland Corners (1842) and so out of Dane Co. (Park 552; Barton 1039) See *The New Janesville Rd.*

Jargo School [ˈdʒɑrˌgo]

For A. O. Jargo and his descendants, who have owned adjoining lands since before 1873. The school is in SE sec. 24, COTTAGE GROVE.

* **Jefferson and Madison Road**

For its termini: an early road, in existence in 1845 (Dane 135), in part the predecessor of present US 18. It passed thru *Buckeye* cor.

* **Jeglum School** [ˈdʒɛgləm]

For the Jeglum family (see *Jeglum Valley*): a former name for *Meadow View School.* (Kitls.) It was in Jeglum Valley until after 1900 (W sec. 33); later it was moved to its present position. (Gay, Blied)

Jeglum Valley

For H. T. Jeglum and his family, Norwegian settlers, who bought land here some time before 1890: the valley chiefly in secs. 32 and 33, PERRY.

Jeraback's Bluff [dʒəˈrɑˌbɛks]

For Anton Jeraback, owner of the land since before 1926: a bluff in NE sec. 6, ROXBURY, next E from *Kehl's Bluff.* (Derl.)

Jimtown, vill.

For Jim Wilson, blacksmith: a nickname for the village of Montrose, NE sec. 30, MONTROSE. This name was locally current from about 1850; when the Montrose PO was established, Wilson tried to get people to give up "Jimtown," but they kept it up all the more. It is still known to the older generation, tho less used than formerly. (Swig.)

Johnson Court, subd.

Because it runs off Johnson St.; within the city of Madison, platted 1902.

* **Johnson School**

Origin uncertain; no Johnsons have lived on land adjoining the school, tho there have been Johnsons nearby since before 1852. A former name for *Kegonsa School.*

* **Johnson's Creek**

For Ole and John Johnson, who from before 1861 owned the land thru which it flows, sec. 36, VIENNA: a former name for *Ella Wheeler Wilcox Cr.* (Ellis)

* **Johnson's School**

For G. Johnson, an early settler (from before 1861 till before

1873) on the land in SW sec. 12, on which this, the first school in SPRINGFIELD, was built (Park 336). Later it was called the *Kingsley School.*

*** Jones Pond**

For D. Jones, who owned the land here from before 1861 till before 1873: a pond in W secs. 7 and 18, RUTLAND; now dried up. (Newton)

Jordahlen Hill [ˈdʒoˑr,dalɛn]

For L. C. Jordahlen, who settled here before 1890, and his descendants: a sharp conical hill where secs. 13, 14, 23, and 24 meet, PLEASANT SPRINGS, and over which the section-line road goes.

*** Junction House, The**

For its location at *Middleton Junction:* a hotel built in 1845 by E. Clewett. Whether the hotel gave its name to the settlement, or the contrary, is uncertain.

*** Junction PO**

For *Middleton Junction.* Est. there Apr. 26, 1870; name changed Jan. 17, 1872, to *East Middleton PO.* See *Middleton PO.*

Juniper Bluff

For the striking patches of juniper on it: a bluff in NE sec. 6, ROXBURY; also known as *Kehl's Bluff.* (Derl.)

Kalkberg [ˈkalk,bɛrç]

German, meaning "lime-hill": a well-known lime quarry, with kilns, on the line between secs. 22 and 23, BERRY. So called in the large local German settlement in early days (i.e., from about 1850). The quarry has not been used for many years, now, but the name is still remembered. (Ketelb.)

Kalscheur's Lake [ˈkalʃəz]

For Cyril Kalscheur, present owner: a small lake (at low water becoming two lakes) in NE sec. 18, SPRINGFIELD. Formerly *Schurtz's, Watzke's,* and *Schroeder's Lake,* as ownership has successively changed.

Kaskeland, settlement [ˈkaskə,land]

A familiar name, among Norwegians, for the *Koshkonong Settlement* (Barton 1009). It is composed of the first two syllables of "Koshkonong," with [ʃ] converted normally by Norwegians to [s], plus the word *land,* which assimilated this to many

Norwegian place-names. Rølvaag uses it in *Giants in the Earth,* 1924 (Haugen); and it was in use much earlier.

* **Keenan School** ['kinən]

For the Keenan family, who gave the land for it, and *Keenan's Cr.*, which flows by behind. A former name for *Oakside School.*

Keenan's Creek

For George Keenan, Irish settler here (1849), and his descendants, owners of land along it: a creek rising in sec. 20, DUNN, and flowing NE into Mud L. on sec. 10.

Kegonsa, cor. [kɪ'gɑnsə]

For the nearby lake: a crossroads settlement between secs. 16 and 21, PLEASANT SPRINGS. Shown on maps from 1911 forward. Also, *Kegonsa Store.*

Kegonsa Grove, subd.

For its location on the wooded SE shore of L. Kegonsa; platted 1896, by Geo. Nicholls, in NW sec. 29, PLEASANT SPRINGS.

Kegonsa Lake

=*Lake Kegonsa.*

Kegonsa Park, subd.

Descriptive; platted 1898, on the wooded shore of L. Kegonsa; SE sec. 30, PLEASANT SPRINGS.

Kegonsa School

For *Kegonsa* (*Store*), and ultimately, nearby L. Kegonsa. At the junction between secs. 16 and 21, PLEASANT SPRINGS. Formerly *Johnson School.*

Kegonsa Store, cor.

For the store, prominent feature of the corner settlement; another name for *Kegonsa,* cor.

Kehl's Bluff [kelz]

For Jacob Kehl, German settler, and his family, who have owned the land from early in the settlement: another name for *Juniper Bluff.* The adjoining bluffs to the E are sometimes referred to, with this, as Kehl's Bluffs, tho the Kehl land does not include them. (Derl.)

* **Kelly Hill**

For S. T. Kelly (or Kelley), who settled here before 1861: a hill in SW sec. 26, SPRINGDALE. The Kelly Hill Cheese Factory is shown here on the Rural map (1910).

Kendall Terrace, subd.

For the nearby Kendall Ave.; within the city of Madison, platted 1928.

Kennedy Pond

For M. Kennedy and his descendants, owners of the land: a small lake or pond in SW sec. 32, WESTPORT.

Kerl School [kɜl]

For Otto Kerl and his descendants, who have owned the land on which it is from before 1861. At the road junction in SW sec. 27, BERRY.

*** Kethel's Hill** ['kɛˌθɛlz]

For James Kethel and his family, who owned the land from before 1861 till after 1904: a hill in the N part of sec. 29, VERONA. (Davids.)

Keyes Springs [kaɪz]

For Elisha W. Keyes, owner of the land from before 1890: some well-known springs near the shore of L. Monona, in NW sec. 17, BLOOMING GROVE. Now within *Shore Acres.*

*** Keyesville**

For its promoter, Joseph Keyes, who built a sawmill there in 1846 (on Koshkonong Cr., a short distance above Cambridge, CHRISTIANA) in the hope of founding a settlement. Since he sold the mill the following year, the name probably gained very little if any currency.

Kilian's Corners ['kɪlɪənz]

For the Kilian brothers, members of an early family, who bought the adjoining land from W. R. Taylor some time between 1899 and 1904: a recent alternate name for *Taylor's Corners.*

Kingsley Corners

See next: the crossroads between secs. 12 and 13, SPRINGFIELD.

Kingsley School

For the Kingsley family (see *Kingsley Corners*, at which the school is). Now closed. Formerly *Johnson's School.*

*** Kingsley Settlement, The**

For Saxton P. Kingsley, of Mass., and his family; he settled, 1856, near what became *Kingsley Corners*, and around his farm grew a settlement which went by his name.

Kittelson Valley ['kɪtl̩sən]

For Ole Kittelson of Telemarken, Norway, and his descendants;

he settled (1854) in SW sec. 28, PERRY: another, more current name for *Pleasant Valley.*

Kittleson Hill

For the Kittleson family, owners of part of the land from before 1926: a hill chiefly on E secs. 26 and 35, COTTAGE GROVE. (Cunn.)

Klevenville, vill. ['klɛvən,vɪl]

For Ivor K. Kleven, an early Norwegian settler at the crossroads nearby to the south and promoter of the village; he had a lumber yard, built the depot (1881), etc. (Gilb.). Formerly platted as *Pine Bluff* village, but the name was changed about 1884; NE sec. 4, SPRINGDALE.

Klevenville PO

For the village. The name was changed from *Bluff PO* Apr. 27, 1891.

Klevenville School

For the village; the school is in NE sec. 4, SPRINGDALE.

Knickerbocker Park, subd.

For the Knickerbocker Ice Co., whose property adjoined; within the city of Madison, platted 1913; secs. 5 and 8, BLOOMING GROVE.

Kohlmanns Buckel ['kol,mɑnz 'bʊkḷ]

= *Kohlman's Hill:* this was the original form of the name (still in existence), which was given and used in the local German farm community, of which Kohlmann had been a member.

Kohlman's Hill ['kolmənz]

For Chas. Kohlmann, a farmer who in 1848 was robbed and murdered there when he was setting out with an oxcart for Milwaukee, to buy supplies for himself and some neighbors: a hill at the junction of secs. 9, 10, 15, and 16, SPRINGFIELD; also *Kohlmanns Buckel, Graveyard Hill.* (The present spelling has dispensed with an original final *n.*)

Koshkonong Creek (formerly, —— **River**) ['kɑʃkə,nɑŋ]

After *Lake Koshkonong,* into which it flows. Shown with this name on maps from 1833 (Surv. T6N,R12E) forward. It rises in the NW part of SUN PRAIRIE, and flows thru COTTAGE GROVE, DEERFIELD, CHRISTIANA, and ALBION; also partly thru Jefferson Co. See also *White Water R.* and *Busseyville Cr.*

*** Koshkonong Ford**

There were two fords of the Koshkonong Creek, one in SE sec. 20, the other in central sec. 34, SUN PRAIRIE. In early days, Winnebago Indian trails crossed here. These were used by the white settlers, and developed into present highway T and a town road. (Eldred)

Koshkonong Lake

=*Lake Koshkonong.*

Koshkonong Prairie(s)

For *Koshkonong Cr.*, the chief nearby stream: a large prairie stretching diagonally thru CHRISTIANA and parts of DEERFIELD, PLEASANT SPRINGS and DUNKIRK. A southerly spur is called *Albion Prairie.*

Koshkonong Settlement, (The)

For the *Koshkonong Prairie*, on which it was made, beginning in 1842: an early, flourishing Norwegian settlement.

*** Krogh's Mill Pond** [krogz]

For Caspar Krogh, who owned the mill from 1848: a former large millpond covering parts of secs. 13, 14, 15, 22, 23, 24, and 27, DEERFIELD, and resulting from the damming of Koshkonong Cr. on sec. 24 about 1840. While it existed, Mud Cr. also flowed into it. When the dam was removed (about 1881–2) the pond was drained, and the land it covered is now in farms.

LaFollette School [ˌla⊥ˈfɑlət]

For Robert M. LaFollette, former Governor of Wisconsin, whose birthplace is just to the N: an alternate but unofficial name for *Dixon School.*

Lagoon du Nord

The last two words are French; thus the name means "North Lagoon": the central channel of *Belle Isle.* This is a "paper" name, put on the plat by L. S. Davis, 1928, but not actually used in the locality. (Owen)

Lagoon du Sud

The last two words are French; thus the name means "South Lagoon": the outmost channel of *Belle Isle.* Named and used like the preceding entry.

Lake Barney

For Barney McGinty, who owned the land on which it is from 1846 until about 1850 (Barry): a small lake on secs. 34 and 35,

FITCHBURG. Tho the name must have been in use from McGinty's time, its first appearance on a map is not till 1904 (USGS).

Lake Drive School

Because the road on which it is (W sec. 35, ALBION) runs down to L. Koshkonong.

Lake Edge Park, subd.

For its location by L. Monona; platted 1909, in SW sec. 9, BLOOMING GROVE. The name was used for a dairy at this location before being given to the plat (Gay 1899).

Lake Forest, subd.

Descriptive; platted 1919, in wooded land along the S side of L. Wingra, in N sec. 34, MADISON.

Lake Harriet (also, **L. Harriett**) ['hærjət]

Origin unknown: a small lake, or pond, in N secs. 8 and 9, OREGON. It is shown from the earliest maps onward, but unnamed before 1861 (Lig.). It is said to have been named for the daughter of D. P. Clark, "the owner of the land on which the lake is largely located" (Brown 7); but since Clark did not own the land till after 1873, this cannot be correct. It may be pertinent that the wife of Harry Brown was named Harriet, and that the Browns settled in sec. 9 in 1848, about half a mile SE of the lake.

Lake Harriet School

For nearby *L. Harriet.* It is in S sec. 5, OREGON, at the crossroads between secs. 5 and 8.

Lake Kegonsa [kɪ'gɑnsə] (formerly, probably also [kɪ'gɑnsɪ])

From an Ojibwa word meaning "little fish." The name was chosen and given this form by Lyman C. Draper, director of the Wis. State Historical Society, in 1854, and was used first on Greeley's map in this year, tho not made official till 1855. It referred to the fact that the Winnebago Indians had considered the outlet of this lake a good fishing ground. The former name (from 1833) was "First Lake"—still in use. "Fish Lake," as the translation of "Kegonsa" given on Greeley's map, may have had some currency. The lake is in DUNN and PLEASANT SPRINGS. See *The Four Lakes.*

Lake Kegonsa Station

For nearby *Lake Kegonsa:* a station on the CMSP railroad, est.

1901, in NW sec. 20, PLEASANT SPRINGS. (Hiway maps show this simply as "Kegonsa.")

Lake Koshkonong ['kɑʃkəˌnɑŋ] (Earlier spellings have been: 1820–2, **Kus-kou-o-nog**; 1828?, **Goosh ke-Hawn**; 1829, **Cos ca hó e nah**; 1833, **Kushkawenong, Kishkanon, Kuskonong**; 1834–5, **Koskonong**; 1835, **Coshconong**; etc. The present spelling was first used by Doty, 1844.)

This name is usually said to mean "the lake we live on," because what is now L. Koshkonong is so labeled on Farmer's map of 1830. However, there is no relation between the two: "Koshkonong" cannot be so translated. This explanation, given by Park, has been widely repeated, tho wrong.

This name appears to be Algonquian, the variant spellings representing the same basic word heard and recorded by different people with differing degrees of accuracy, and probably from Indians speaking different dialects. "Koshkonong" is a regular contraction of the longer forms, of which "Kushkawenong" represents Ojibwa, "Cos ca hó e nah" and "Goosh ke-Hawn" probably Menomini, and "Kus-kou-o-nog" Potawatomi or Fox. The meaning may be rendered as *Where there is heavy fog*, or *Where it is closed in by fog*. (Geary, Lincn.)

It should be noted that the early applications of the name are not limited to the lake. Morse (1820–2) reports that it was the name of the largest Winnebago village in the Rock R. valley, and of the lake on which the village was (now L. Koshkonong). This implies extension from the village to the lake.

Farmer (1828?) and Chandler (1829) apply the name on their maps to what is now the Yahara R., which enters the Rock R. a short way below L. Koshkonong. (The present Koshkonong Cr. was not named on their maps, and was not given this name on a map till 1833.) Thus "Koshkonong" was probably applied generally to a region, and specifically to various features (village, lake, creek) within that region (cf. *Taychopera*). The meaning of the name suits such a general application. The authority of the Government survey of 1833–6 seems to have stabilized the application of "Koshkonong" to the lake, and thence it has spread to the present creek, the prairie, the settlement, and various establishments in the settlement. Although the name is Algonquian, it was applied to a Winnebago (Siouan) village.

This probably means no more than that the information on this region came to the white man via the Algonquian tribes, rather than directly from the Winnebago. The Winnebago here were surrounded by Algonquians, who passed thru this territory frequently. The word "Winnebago" itself is an Algonquian word, tho used regularly of a non-Algonquian tribe—an exactly parallel example.

Lakeland, subd. (or, **Lake Land**) ['lek,lænd]

Descriptive (L. Mendota is not far off); platted 1855, in W sec. 21, MADISON.

Lake Lawn, subd. ['lek,lɔn]

Descriptive; platted 1869, within the city of Madison and near to L. Mendota; sec. 14, MADISON.

Lake Marion ['mærɪən, 'mærjən]

So named by Frank Murrish, of Mazomanie, because his aunt, Frank [*sic*] Marion, visiting from Chicago, fell into the lake. (Pick.) The "lake" is really a millpond (sec. 16, MAZOMANIE), also locally called *Upper Pond.* It was formed in 1856, but the name Lake Marion does not appear on a map till 1890 (Foote).

Lake Mendota [mɛn'dotə] (formerly, probably also [mɛn'dotɪ])

Probably for the village of Mendota, Dakota Co., Minn. The word is Dakota (*mdó-te*) meaning a confluence of rivers, in which sense it applies exactly to the location of the Minnesota village. It was suggested as the name for this lake (which covers parts of WESTPORT, MIDDLETON, and MADISON) in 1849, by Frank Hudson, of Madison, who chose it for its euphony and its association with Indian legends. It became official in 1855. The former name was *Fourth L.; Great L.*, a supposed translation of "Mendota," also had some currency. See also *The Four Lakes*, and *Taychopera.*

Lake Monona [mə'nonə] (formerly, probably also [mə'nonɪ])

Probably for the town of Monona, Clayton Co., Iowa. The origin of the word is uncertain, but it may be Sauk-Fox, the name of a beneficent female deity. The application of the name to the Wisconsin lake was made at the suggestion of Frank Hudson, of Madison, in 1849, and became official in 1855; Hudson chose it for its euphony and its association with Indian legends. The former name was *Third Lake; Fairy Lake*, given on Greeley's map (1854) as a translation of "Monona," may also have had some currency. The lake covers parts of MADISON and BLOOMING GROVE. See *The Four Lakes.*

Lake Park, subd.

Because formed out of a park (*Schuetzen Park*), and bordering on L. Monona; platted 1906, in NE sec. 7, BLOOMING GROVE; since absorbed into the city of Madison.

Lake Park Addition, subd.

Because it was made in the NE part of the village of Mount Horeb, in the direction of a park containing an artificial lake, which had been recently opened; platted 1914, in NE sec. 11, BLUE MOUNDS.

Lakeside, subd.

For former local establishments, the Lake Side Water Cure resort, begun 1854, and the Lakeside House hotel, which burned 1877; and ultimately for their location by L. Monona. Platted 1901 as an addition to South Madison; NE sec. 26, MADISON.

Lakeside School

For its location near L. Kegonsa. In NE sec. 27, DUNN. Formerly *Colladay School*, then *Greene School.*

Lake View, subd.

Descriptive (L. Mendota may be seen from here); platted 1898, in SW sec. 18, MADISON.

Lakeview, cor.

For the PO; a small crossroads settlement in sec. 24, FITCHBURG.

Lake View Heights, subd.

Descriptive (L. Mendota may be seen from here); platted 1940, in SE sec. 25, WESTPORT.

Lake View Place, subd.

Descriptive (L. Monona may be seen from here); within the city of Madison; platted 1930.

* **Lake View PO** (later, **Lakeview PO**)

For its point of location (at the crossroads, sec. 24, FITCHBURG) from which L. Waubesa is visible. Est. July 1, 1848, disct. Mar. 21, 1855; re-est. Jan. 29, 1856, disct. July 21, 1864; re-est. May 20, 1872, disct. June 20, 1887; re-est. July 14, 1887, disct. May 31, 1901. The spelling was changed to "Lakeview," June 28, 1895.

Lakeview Road, The

A former alternate name for the *Oregon Rd.:* now falling into disuse.

Lakeview School

For nearby *Lakeview.* S sec. 24, FITCHBURG. Now closed.

*** Lakeville**

Error (Lapham 1849) for *Lakeview.*

*** Lake Washington**

For William Washington Patrick (Barton); see *Patrick's Lake,* of which it was a local alternate, now apparently forgotten.

Lake Waubesa [ˌwɔ'bisə] (formerly, probably also [ˌwɔ'bisɪ])

From Ojibwa, meaning "swan." The name was chosen and given this form by Lyman W. Draper, first director of the Wis. State Historical Society, in 1854, and was used first on Greeley's map in this year, tho not made official till 1855. It referred to the fact that an unusually large swan had once been shot here. The former name was *Second Lake; Swan Lake,* as the translation of "Waubesa" given on Greeley's map, may also have had some currency. The lake covers parts of BLOOMING GROVE and DUNN. See *The Four Lakes.*

Lake Waubesa Station

For nearby *Lake Waubesa:* a station on the CMSP railroad, est. 1901, between secs. 33 and 34, BLOOMING GROVE. (USGS and Hiway maps show this simply as "Waubesa.")

Lake Wingra ['wɪŋgrə, 'wɪngrə]

From Winnebago ['wĩçrɑ], meaning "duck," this lake having been, in early times, a great resort of ducks. The lake was shown on many maps from 1835 (Terr.) forward, but without name. J. A. Noonan, who came to seek land in the region in 1837, claimed (WHS *Coll.,* 7.410) to be the first American to have learned this name "from Joe Pelkie, the early French settler," and to have had it placed on a map "in that month of February, by Heading and Delaplaine, of Milwaukee." This map has not been found (it must have been a MS., since no maps were being printed in Milwaukee in 1837), but it is at least certain that Heading, formerly associated with Delaplaine, was engaged in map-making in Milwaukee in that year (see WHS *Coll.,* 11.245). Certainly it was on paper that year, when the owners of the *Western Addition to Madison* had plats for this subdivision recorded.

The name first appears in print on two maps of 1839: Cram's had "Weengra Lake" erroneously applied to *Hook Lake,* although Wingra Lake is drawn in correctly, without name; Hagner's, made under Cram's direction, shows the name spelled

"Wingra L.," and correctly applied. Perhaps the error on Cram's map may thus be laid to the engraver. Doty (1844) preserves Cram's spelling, but most later maps use Hagner's spelling, which has become accepted.

Noonan says not only that Wingra meant "duck," but that it "was the Indian name for that body of water"; presumably, then, "Duck Lake," which was another early name, was a translation.

Still another early name was "Dead Lake," apparently given by the whites, in reference to the supposed lack of an outlet, since the lake seemed to merge into swamp, though today *Wingra Creek*, or *Murphy's Creek* connects it with L. Monona. It is noticeable that such a map as Morrison's (1837) shows the lake completely surrounded by swamp; only one map (Hagner's) before 1846 shows an out-flowing creek. Even the USGS map (1904) shows no direct connection: "Murphy Creek" rises in the swamp surrounding the lake. Noonan used the phrase "the Dead Lake—more properly Duck or Wingra Lake," which indicates perhaps the relative early popularity of the names. The third name won the competition, one may guess, because of its exotic or individual quality, and perhaps because it suited well with the names of the adjoining *Four Lakes*. But "Dead Lake" was still in use late in the 19th century. See also *The Little Lake*.

The lake covers parts of secs. 26, 27, and 28, MADISON.

Lakewood, subd.

Descriptive and probably reminiscent; in wooded land along the E shore of L. Mendota, in SE sec. 1, MADISON; platted 1912, since absorbed into *Maple Bluff* village. It was owned by the Lakewood Land Co. for several years before being platted.

*** Lakewood Bluff,** vill.

A proposed name for the village which was in the process of being formed, in 1931, out of *Lakewood* and *Maple Bluff* plats, and others—the name to be combined as the plats were. It is shown in Thrift's platbook, tho it never came into being; the name chosen instead was *Maple Bluff* village.

Lansing Place, subd.

For W. H. and Martha Lansing, former owners of the land; platted 1926, in SE sec. 5, BLOOMING GROVE.

Lappey School ['læplɪ]

For F. W. Lappley, who has owned land nearby since before 1926: a former name for *Pleasant Site School*.

*** Lappley's Corners**
For John Lappley, who owned the land surrounding it from before 1890 till after 1899: a former name for *Nonn's Corners;* it superseded *Schaefer's Corners.*

Larson's Beach
For J. O. Larson, owner since before 1911: a section of the shore of L. Waubesa in SW sec. 34, BLOOMING GROVE.

Lawrence Hill
For the Lawrence family, who owned adjoining land from the 1840's until about 1930: a local alternate name for *Token Creek Hill.*

Laws' Landing
For Moses Laws, who started a ferry in 1855 between Ferry Bluff (Sauk Co.) and this place: a point of land within a bend of the Wisconsin R., in SE sec. 20, MAZOMANIE. The name is still in use tho the ferry was discontinued long ago. (Laws, Derl.)

*** Lawton's Corners**
For G. F. Lawton and his family, who owned land here from before 1861 till after 1911: the junction between secs. 17 and 20, ALBION.

*** Lead Road, The** (or, —— **Route**) [lɛd]
Because lead was hauled over it by ox-team in the late 1830's and the 1840's from Mineral Point to Milwaukee. In Dane Co. it apparently left the Military Road at some point E of Blue Mounds and turned SE across *Stoner's Prairie* to *Runey's Tavern,* thence E thru RUTLAND and NE past *Mayhew's.* (Park 355, 448, 505, 513)

Lee Creek
For N. E. Lee and family: tributary of the Pecatonica R., running thru *Lee Valley.* (Kitls.)

*** Lee School**
For Nels A. Lee, of Norway, and his descendants, who have owned land adjoining Liberty Corners to the S, from before 1873, and on which the school has been (see *Prescott School*). This is a former name for *Liberty School.*

Lee's Park, subd.
For Peter and Carrie Lee, owners of the land; on L. Kegonsa, platted 1921; SW sec. 18, PLEASANT SPRINGS.

Lee Valley

For N. E. Lee (who settled here before 1873) and his descendants: the valley, chiefly in sec. 34, PERRY, thru which *Lee Creek* runs. (Kitls.)

*** Leicester PO** (also, **Lester PO**) ['lɛstɚ]

For Leicestershire, England, from which many of the early settlers of the region, including the postmaster, William Crow, came. (See Park, 564) Est. Apr. 8, 1852, and kept in NW sec. 5, WESTPORT. Disct. May 18, 1853; re-est. May 31, 1855. The name was changed to *Waunakee*, and it was moved to that new village, Oct. 11, 1871.

Lenaas School ['linɑˏs]

For the Lenaas family, Norwegian settlers. A. A. and Ole B. Lenaas owned land in this section from before 1873 till after 1911. The school is in SW sec. 7, CHRISTIANA.

*** Lester PO**

=*Leicester PO;* a phonetic spelling used on some maps; unofficial.

Liberty Corners

On Liberty Prairie, and for Liberty Church, just N: a crossroad at the corners of secs. 29, 30, 31, and 32, DEERFIELD.

Liberty Hill

See *Liberty Prairie:* an isolated conical hill, mostly in NW sec. 2, PLEASANT SPRINGS. Also called *Liberty Mound.*

Liberty Mound

=*Liberty Hill;* the less common form.

Liberty Prairie

So named in the early days of the settlement (i.e., in the early 1840's) by a company of Fourth-of-July celebrators, who, it is said, climbed up on top of *Liberty Hill* (named on the same occasion), viewed the prairie below, and (with the aid of a little brown jug) were inspired to give these appellations. The prairie covers parts of secs. 33, 34, 35, and 36, SE in COTTAGE GROVE, and 31 and 32, SW in the adjoining DEERFIELD.

Liberty Prairie Church

For its location : a Norwegian Lutheran church; built 1851, and in continuous use since; SW sec. 29, DEERFIELD.

Liberty School (also, **Liberty Corners ——**)

For *Liberty Prairie* and *Liberty Corners.* In SW sec. 29, DEERFIELD. Formerly also called *Lee School* and *Prescott School.*

Lime Kiln Spring(s)

For a lime kiln located near it in early days: a spring south of Monroe St., now in the University Arboretum; NE sec. 28, MADISON. The name was in use as early as 1844 (Dane 111). Also known as *Marston Spring*. (Brown 7)

Lincoln Park, subd.

For Abraham Lincoln? (There is an earlier *Washington Park* north along the lake shore.) On L. Kegonsa; platted 1905, in N sec. 26, DUNN.

Linden Hill, subd.

Descriptive ("linden," however, is a literary name for the tree called "basswood" in this region): an addition to *Fairoaks*, platted 1914, and later absorbed with it into the city of Madison. Sec. 5, BLOOMING GROVE.

Little Door Creek

By contrast with *Big Door Cr.* Since this latter is now officially called Door Cr., Little Door Cr. is considered a tributary. It rises S in COTTAGE GROVE, and flows N and W to join Door Cr. on sec. 32. Used on maps as late as 1905 (USGS); still used locally.

*** Little Lake, The**

Descriptive, by contrast with Lakes Mendota and Monona: an early name for *L. Wingra*. Used by Durrie, 50 (1874).

Little Norway

Descriptive: a farm on NW sec. 4, BLUE MOUNDS, whose buildings and equipment are of Norwegian style, original and in reproduction—a sort of museum. Its other name is *Nissedahle*.

Little Sugar River

As a small but important branch of the Sugar R. It rises from several sources in S PRIMROSE, flowing at once into Green Co., SE, then S, then E to join Sugar R. in ALBANY. Shown on maps from 1832 (Surv. T3N,R7E) onward. (Some recent Highway maps have erroneously labeled the West Branch of Sugar R. with this name.)

Little Valley School

Descriptive of its location: it is in a valley between considerable bluffs, in SE sec. 5, ROXBURY. Now closed.

Livesey's Spring ['lɪvˌziz]

For James Livesey and his descendants, owners from before 1861 of the land (SW sec. 6, MADISON) on which it rose: a spring early important in the history of the Pheasant Branch region, since Michel St. Cyr, first local settler, had his fur-trading post near by, and many travelers knew it. Formerly called *Belle Fountaine*. Now filled in, tho there are other springs nearby.

Livesey's Woods

For James Livesey and family (see preceding entry): the heavily wooded land mostly in sec. 6, MADISON. (Park 555)

Livick's Pond ['laɪvɪks]

For I. Livick and his family, who, from before 1873 till after 1904, owned the land on which the pond is (NW sec. 36, DUNKIRK).

London, vill.

For its settlement by natives of London, England (Sten.): a village, platted 1882, sec. 25, DEERFIELD. Probably so named by its platter, Archibald Armstrong—of English descent, tho from N Ireland. (Thor.)

London PO

For the village. Est. Apr. 27, 1882.

Lone Oak School

Subjectively descriptive, in part, tho there is a single oak close by. In SE sec. 34, YORK.

*** Long Breaking, The**

Presumably a long piece of "broken" or plowed land, in the N part of MAZOMANIE; so called by the settlers about 1845 (Park 267).

Lorch School ['lɔrtʃ]

For John Lorch, of Prussia, who settled in *Pheasant Branch* village in 1851, and became a prominent storekeeper. The school is in the village.

Lost Lake

A name of Indian legendary origin, used in many places in the USA; applied to *Kennedy Pond* (rather appropriately, since it was then surrounded with thick woods) by a visiting New York woman writer (Corc.). The name has been widely adopted in the locality.

Lost Village, The

For its being rather isolated from Madison: that part of the Lake Forest plat which was built up (N sec. 34, MADISON). The plat was an ambitious one, and it did not develop as much as was expected. A local name.

Lotus Lake

Because it was once thickly grown with yellow pond-lilies or "lotuses": a local unofficial name for *Turtle L.*

Lower Hoboken, beach

Because it is below *Hoboken Beach,* closer to the outlet of L. Monona; it is shown on Rural map (1910) in SW sec. 20, BLOOMING GROVE.

Lower Mud Lake

By contrast with *Upper Mud L.;* an alternative local name for *Mud L.*, DUNN.

Lower Pond, The

By reference to the Upper Pond (Lake Marion): a small pond at a lower level along Black Earth Cr.; secs. 8 and 9, MAZOMANIE. (Linc.)

Luhman's Corners ['lumənz]

For August Luhman, Jr., storekeeper there, and his descendants: the crossroad formerly called *Brackenwagen's.* Luhman succeeded Brackenwagen in the 1890's.

Lukken School ['lʌkən]

For Ole Lukken, owner of the land (from before 1904) on which the school is: on the town road in NW sec. 26, BLUE MOUNDS.

Lund's Point [lʌndz]

For Mr. Lund, owner: a piece of land jutting into L. Kegonsa, in SW sec. 25, DUNN. Formerly *Quam's Point,* and before that, *Cedar Point.*

*** Lutheran Hill** ['luθɚən]

For the settlement of Lutherans hereabouts, and *St. John's Lutheran Church* near the hilltop: a local name for *Springfield Hill,* found only on USGS map, 1906–7, and Soils.

*** Lyle PO** [lɑɪl]

For John Lyle, Scottish settler, owner of large tracts of land in secs. 17 and 18, MONTROSE, on which the PO was kept. Est. Apr. 29, 1892; disc. Dec. 15, 1900. "Lyle" is still put on platbooks, tho there is no settlement.

Lyle School

For the former *Lyle PO* that was near here; E sec. 18, MONTROSE.

*** Lynch's Corner**

For Peter Lynch, who, in the 1890's, owned land adjoining the corner in SW sec. 5, VERMONT, at which *Peculiar PO* was kept. (Mick.)

* **Macbride's Point** (also, **McBride's** ——)

For James D. Macbride, who came to Madison in 1842–3, and bought the land including this point. This name was in use at least up thru the 1880's; it was also called *Maple Bluff*, its name today, and for a time, *Stone Quarry Point*, and *Eagle Point*.

* **Macbride's Point Landing**

A steamboat landing at *Macbride's Point* in L. Mendota. Used in the 1870's and 1880's.

Macfarland, vill. (also, **McFarland, MacFarland**) [mək'fɑrlənd]

For Wm. H. McFarland, original owner of the land. The settlement began when McFarland built in sec. 3, DUNN, for the CMSP railroad, a depot which was given his name, 1856; he platted the village the same year; a PO was est. there 1857. It was incorporated 1920, and includes parts of secs. 2, 3, and 4.

Usage is divided in the spelling of the name. The PO and incorporated village do not capitalize the F; maps and county histories do capitalize it. The latter spelling is, of course, older. See *City of the Second L.*

Macfarland PO

For the village. Est. June 1, 1857.

* **Mackessey's Corner** ['mækɪsɪz]

For Mrs. T. Mackessey, who owned the surrounding land from before 1904 until after 1931: the junction in SE sec. 33, CROSS PLAINS. (Dahmen)

Madigan School ['mædɪgən]

For T. Madigan and his descendants, who have owned adjoining land since before 1873: at the road junction, in NW sec. 26, DANE. Now closed.

Madison, city ['mædɪsən, 'mædɪsn̩]

For James Madison, fourth President of the US, who had just died on June 24 of the year of the naming, and who had been admired by Judge J. D. Doty, the projector of the city. Doty planned to make this the capital of the State (then the Territory) of Wisconsin; he accordingly had it surveyed and platted in mid-November, 1836, and it was accepted by the legislative act of Nov. 28. (For further details, see WHS *Coll.* 6.388–96.) The first map to bear the name is the surveyor's plat (Suydam, 1836).

This is dated at Green Bay, Oct. 27, 1836, which must be an error: all other evidence points to Nov. as the month. Besides, the title includes the phrase "Madison, the Capital of Wisconsin," which it did not become till Nov. 28. The date on this plat must therefore have been set back by the publisher or someone else.

The hint by Butler (86, note) that Doty was seeking to honor Dolly Madison, in giving this name, is probably wrong. There is no evidence that Doty met the ex-President's widow until after he went to Washington as a Territorial Delegate, 1838. (Smith A)

The original plat is bounded by L. Mendota (N), the Yahara R. (E), Lake Monona (S), and a line one block W of Bedford St., running from L. Monona to the E line of secs. 15 and 22 (W). Madison became an incorporated village, Feb. 3, 1846, and an incorporated city, Mar. 7, 1856. Formerly also called *Taychopera.*

Madison, town

For the village (now city) of Madison, chief settlement in the town. It was organized Feb. 2, 1846, and then included all of Dane Co. except the present ALBION, BRISTOL, CHRISTIANA, DUNKIRK, DUNN, FITCHBURG, MEDINA, OREGON, RUTLAND, SUN PRAIRIE, and YORK. The gradual separation of other towns, however, reduced it, by 1859, to its present outer boundaries, T7N,R9E. Growth of villages and of the city of Madison is still reducing it internally.

*** Madison and Portage Railroad**

For the main towns along it: a former name (1870–1) for a branch of the CMSP railroad. It passes thru Windsor, De Forest, and Morrisonville, leaving Dane Co. from sec. 1, VIENNA. (Barton 794)

*** Madison Branch Road, The**

Descriptive: the local name (in 1877: Park 231) for the branch of the CMSP railroad passing thru Marshall and Sun Prairie to Madison.

Madison Creek

For the city of Madison; so called in the Sauk City region, because the creek is the first which must be crossed on the way to Madison (i.e., by U. S. highway 12, on sec. 18, ROXBURY). Only the upper part of the creek is so called. Formerly, *Inama's Cr.* (Derl.)

Madison PO

For the village, now city. Est. Feb. 3, 1837.

*** Madison Precinct**

For the then village of Madison, its center: an election precinct est. May 16, 1839. (Dane)

Madison Square, subd.

For its location within the city of Madison, but probably also reminiscent; platted 1903, in sec. 6, BLOOMING GROVE.

*** Madison, Sun Prairie, and Watertown Railroad**

For the main towns along it: a former name for a branch of the CMSP railroad. It was completed to Sun Prairie in 1859, to Madison in 1868–9. It enters Dane Co. on sec. 12, MEDINA.

Magelsen School [ˈmɑgl̩sən]

For J. W. Magelsen, who had an orchard adjoining the school, from before 1873 till before 1890. At the junction where secs. 23, 24, 25, and 26 meet, PLEASANT SPRINGS. Formerly *Huber School.*

*** Maine Settlement, The**

For the state of the settlers' origin: an early settlement in the N part of RUTLAND (Durrie); begun about 1845 (Park 388).

*** Malloy School** [məˈlɔɪ]

For Michael Malloy, who bought land nearby in sec. 6, before 1873: a former name for *Schaller School.*

Malone School [məˈlon]

For Axium Malone; see *Malone Valley.* The school is in SW sec. 29, SPRINGDALE.

Malone Valley

For Axium Malone and his descendants, who owned adjoining land from 1846 till after 1931: the valley of the W branch of Deer Cr., particularly in secs. 19 and 29, SPRINGDALE. Malone, despite his name, was Pennsylvania Dutch. (Gilb.)

*** Manchester**

Origin unknown; possibly reminiscent (there were already 10 other Manchesters in the US), or for somebody by that name: a "paper town," platted Jan. 12, 1837, in SW sec. 28, and shown on early maps on the E side of the Yahara R., at the confluence with the creek flowing from the big spring, PLEASANT SPRINGS.

*** Mandamus**

From Latin, "we command"—a legal term, for a writ to enforce obedience. How or why it came to be the name of this "paper town" is unknown. Tho no plat was recorded for it, it is shown on early maps (Terr. 1835 and others as late as 1840) at or just S of the present village of *Pheasant Branch*, which superseded it, and preserved its name in that of a street on the village plat—which street (consistently enough) never was built.

Maple Bluff

Descriptive: a high promontory projecting into L. Mendota, in sec. 1, MADISON, on which stood, in early days "the finest sugar maple grove in the territory" (Stoner). The Govt. surveyor, Orson Lyon, shows here on his notebook map (1834) "Sugar Grove." But how early the name "Maple Bluff" was in use is a question. The known early name was *Macbride's Point; Maple Bluff* was in existence at least as early as 1880, since Butt. uses it; it appears on a map first in 1890 (Foote).

Maple Bluff, subd. For *Maple Bluff*, at which it is; platted 1927, in NW sec. 1, MADISON.

Maple Bluff, vill.

Ultimately, for the promontory so called, but proximately for the Maple Bluff plat, which, with others, went to the formation of the village; incorporated 1931, and including much of sec. 1 BLOOMING GROVE.

Maple Center School

For some nearby maples; the "center" is perhaps a correspondence with nearby York Center School. At the crossroads where secs. 20, 21, 28, and 29 meet, YORK.

Maple Corners School

For the maple trees nearby. At the road junction in S sec. 30, FITCHBURG.

Maple Court, subd.

From its relation to Maple St., in the city of Madison; platted 1896; SW sec. 26, MADISON, about a central street named Maple Court, an extension of Maple (now Gilson) St. The name was probably arbitrary, since the parallel streets were also named for trees: Spruce, Cedar, and Pine. Since absorbed into the city of Madison.

Maple Grove School (ALBION)

Descriptive: it is among maple trees (NE sec. 8).

Maple Grove School (BRISTOL)
For the nearby maple trees: in SE sec. 6.
Maple Grove School (VERONA)
For the nearby maple trees: in NW sec. 12.
Maple Grove School (WINDSOR)
For its location among maple trees: in SW sec. 25.
Maple Knoll School
For its location on rising ground, among maple trees: in SE sec. 26, BRISTOL.
Maple Lane
Descriptive, because it is lined on both sides with large maples: a part of the road running N from county trunk B for about a quarter of a mile in N sec. 21, DUNN.
Maple Lawn Heights, subd.
For Maple Lawn Farm, the descriptive name of the land before it was platted (1915); in NE sec. 3, FITCHBURG.
Marlborough Heights, subd. [ˈmɑrlˌbʌrə]
Origin unknown; probably chosen as suggestive of distinction; platted 1918, in SW sec. 32, MADISON.
Marshall, vill.
For Samuel Marshall, who, with W. H. Porter, bought the property from A. M. Hanchett. The village includes parts of secs. 10 and 15, MEDINA. It was incorporated 1905. The successive names have been *Bird's Ruins* (1839), *Hanchettsville* (1849), *Howard City* (1857), and *Marshall* (shortly after 1861).
Marshall PO
For the village. Changed from Hanchettsville PO, Feb. 19, 1862.
Marsh Creek
Descriptive: it flows from *Hesseltine Marsh*, MAZOMANIE, and westward from sec. 4, into the Wisconsin R.
Marsh Valley
Descriptive. A valley, chiefly in secs. 14 and 15, MAZOMANIE, thru which flows a tributary to Halfway Prairie Creek. (USGS 1916)
Marston Spring [ˈmɑrstən]
For J. T. Marston, who owned the land it was on from about 1870 until the early 1900's: a spring south of Monroe St., now in the University Arboretum; sec. 28, MADISON. Also known as *Lime Kiln Spring*. (Brown 7)

Martinsville, vill. ['mɑrtɪnzˌvɪl, 'mɑrtn̩zˌvɪl]

For *St. Martin's Church,* here. The settlement had existed without name from about the 1860's. This name must have come in with the building of the church; it certainly existed before 1900, when *Martinsville PO* was established here.

*** Martinville PO**

For *Martinsville.* It is likely that the form of the name used for the PO was a simplification made by the US Postal Dept., in line with current policies. Est. Aug. 31, 1900; disct. Jan. 15, 1903.

Marvin Park, subd.

For Henry H. Marvin, of Ohio, owner of the land; platted 1899, added to the vill. of Oregon; NE sec. 11 and NW sec. 12, OREGON.

Marxville, vill. ['marksˌvɪl]

For Johannes Marx, owner of considerable land at and near the crossroads settlement at the junction of secs. 3, 4, 9, and 10, BERRY, the site of the village. Marx, tho he came before 1860, was not one of the earliest settlers; so there was some resentment when the PO was established with his name. It was felt that it should have been *Myer's Corners,* and that Marx's friendship with the town chairman had unduly influenced the decision. Marx is said to have reconciled things by setting up a barrel of beer for the villagers' celebration (Walser). The naming of the village was thus concurrent with that of the PO.

*** Marxville PO**

For Johannes Marx; see *Marxville,* at which it was kept. Est. Apr. 21, 1886; disct. Sept. 30, 1902.

Marxville School

For the village. The school is in SE sec. 3, BERRY.

Maticwaup, subd. ['mætɪkˌwɔp]

Of uncertain origin; the location of the plat on L. Monona makes it possible that the name may be from Ojibwa *mātikwāpi,* meaning "a gathering place for getting water," "the best place for getting water," or the like (Geary, Lincn.). Platted (1906) by T. H. and Bessie G. Brand, who probably named it; NE sec. 26, MADISON; since absorbed into the city of Madison.

Matz's Hill ['mætsɪz, 'mætsəz]

For Ervin Matz, owner of the land since about the 1920's: a prominent, almost isolated bluff in SE sec. 5, BERRY.

Mauneeshah River (also, **Maunesha Creek**) ['mɔni,ʃɔ]

A distortion of, or mistake for, *Nauneesha R.;* however, it still has some currency (tho it is unofficial, and *Waterloo Cr.* official), whereas "Nauneesha" is virtually unknown.

* **Mayhew's,** cor. ['me,hɪuz]

For Wm. M. Mayhew, a Southerner, first settler of the town, who, in 1837, built a log house and tavern at the main crossroads in what is now NE sec. 28, CHRISTIANA. This was a great resort of teamsters, since the *Lead Road* passed thru it, and tho no real settlement grew here, it is shown on early maps (Lapham 1849, etc.). Mayhew was the first postmaster of the town (1846–9), and chairman of the town board.

Maywood, subd.

Formed from the names of A. L. May and W. H. Woodward, law partners; the latter was one of the owners and platters (Woodw.). Within the village of Monona, platted 1939.

Mazomanie, town [,mezo'menɪ]

For the village. The town was separately organized out of the then BLACK EARTH and part of ROXBURY, Jan. 15, 1859; it thus included the N half of T8N,R6E and the fractional T9N,R6E lying S of the Wis. River, except secs. 13, 24, 25, 35, and 36. In 1865 (Feb. 27), ROXBURY yielded these last sections, bringing Mazomanie to its present area.

Mazomanie, vill. (also, formerly, **Mazo Manie, Mazo-Manie, Mazomania;** currently abbreviated colloquially to **Mazo**)

The name was given, and put into its present form, by Edward Brodhead of N. Y., who, with others, platted the village in 1855. It is from the name of the Winnebago Indian, usually spelt "Manzemoneka." On Oct. 18, 1836, this Indian had killed Pierre Paquette, a widely known trader, at Fort Winnebago (now Portage). He was convicted of murder, retried and acquitted, and the whole case made a great stir at the time. Nineteen years later, when Brodhead was "booming" the new village and seeking a name for it, he heard the story of Manzemoneka, was impressed with it, and simplified the name of the Indian to make that of the village. (See Kittle 57–8)

The name has two variant forms, depending on whether the nasalization of the first vowel has been represented in the spelling —in other words, on whether the spellings have an *n* or not:

with *n*, *Manzemoneka* (WHS *Coll.* VII, 356), *Manzamonekah* (*Ibid.* 387); without *n*, *Mahzahmahneekah* (*Ibid.* VIII, 318), *Mauzemoneka* (*Ibid.* XII, 402), *Mauzamoneka* (*Ibid.* XIV, 80–1). "Mazomanie" must clearly have been made from the *n*-less form by omission of the last syllable and respelling of the rest in a more English way. The original accentuation was not changed.

The name means, in Winnebago, "iron walker" or "he who walks on iron," and it may well be that this partly influenced Brodhead to choose it; since it was in anticipation of the coming of the railroad (which reached the village in 1856) that he platted and named it so, he may have interpreted "iron walker" to refer to the steam engine. (Earlier derivations of the name from Ojibwa, meaning "moose-berries," etc., are erroneous.)

The village was originally platted in NW sec. 16 only, but now includes adjoining parts of secs. 8 and 9, MAZOMANIE. It was incorporated 1885.

Mazomanie PO

For the village. When it was est., Jan. 4, 1856, the name was spelled "Mazo Manie"; it was changed to "Mazomanie" Nov. 27, 1895, tho this form had been used for the village as early as 1861.

McComb's Corners [mə'komz]

For Robert McComb (who settled here before 1848) and his descendants, owners of the adjoining land till the early 1930's: the crossroads where secs. 6 and 7, DUNKIRK, and 1 and 12, RUTLAND meet.

McCord Rock(s) [mə'kɔrd]

For E. K. McCord, who owned the land from before 1861 till before 1873: an earlier name for the *Donald Rock*, now very little used.

McFadden Spring(s) [mək'fædn̩]

= *Grand Springs;* the land was bought, 1841, by George McFadden, the earliest local settler, and around his tavern and barn the first settlement sprang up. The name is remembered by the older residents, but is virtually out of use; see *Remy Spring*.

McKenna Park, subd. [mə'kɛnə]

For the McKenna family, an early one in this locality, and specifically Harry McKenna, who had this platted, 1911, in NW sec. 31, BURKE.

McPherson School (mək'fɝsn̩]

For James P. McPherson, Scottish settler, who from 1850 till the 1890's owned the land on which it is: sec. 25, SPRINGDALE.

Meadow Brook Spring

Descriptive: it is the source of a small stream, flowing formerly thru a meadow, into L. Wingra; sec. 28, MADISON. (Brown 7)

Meadow View School

For its location, from which some level land may be seen among surrounding bluffs: S sec. 32, PERRY. Formerly *Jeglum School.*

Medina, town [mɪ'daɪnə, mə'daɪnə]

Said to be for Medina Co., Ohio, because many of the early settlers came from there. In the biographical section of Butt., however, there is only one person still listed (1880) as from Ohio: Joseph Wilt, born in Alsace, France; brought to Medina Co., Ohio, a boy of 14, where he grew up; settled here 1843 at the age of 26. Most of the other settlers were from N. Y. state (in which there is a Medina, Orleans Co.), so perhaps Medina, Ohio, was not the only source of the name. Est. Mar. 11, 1848, by separation from SUN PRAIRIE: T8N,R12E.

Medina School

For the town. At the crossroads in the middle of sec. 32, MEDINA.

Meffert School ['mɛfɚt]

For John Meffert and his descendants, who have owned the land since before 1890. The school is on the N edge of sec. 22, SPRINGFIELD. Now closed. Formerly called *Stone College.*

Meier School ['maɪɚ]

For the Meier family, whose members have owned adjoining land from before 1861. The school is on the town road in NW sec. 24, BLOOMING GROVE.

Meyer's Corners

=*Myer's Corners*

* **Mendota,** vill. [mɛn'dotə]

For the PO; one of the early names of present *Middleton.* The name was abandoned, according to Butt., when "a Mendota started up in another locality"; he probably refers to *Mendota Station*, WESTPORT.

Mendota Beach, subd.

For its position along L. Mendota; platted 1896, in sec. 18, MADISON.

Mendota Beach Heights, subd.

For its position above *Mendota Beach;* platted 1929, in NW sec. 18, MADISON.

*** Mendota Beach Station**

For its position near *Mendota Beach:* a station on the CMSP railroad, in SW sec. 18, MADISON. It is not listed in the Railway Guides, but USGS map (1904) shows it in existence then.

Mendota Court, subd.

For its proximity to L. Mendota; platted 1889, within the city of Madison; SW sec. 14, MADISON.

Mendota Heights, subd.

Descriptive; platted 1890, on high land along L. Mendota, in NE sec. 17, MADISON. Since absorbed into *Shorewood Hills.*

Mendota Lake

=*Lake Mendota.*

*** Mendota Lake Ridge**

For its location with respect to L. Mendota, tho just where it was in unknown. It is mentioned by Park, 17, and was apparently at the E end of the lake. No longer in use.

*** Mendota PO**

For nearby L. Mendota. The name was changed to this from *Middleton Station PO,* July 30, 1862; this was discontinued in favor of *Middleton PO,* newly moved to this location.

Mendota Station (also, **Mendota**)

For Mendota Hospital, and ultimately L. Mendota: a station on the C and NW railroad, SE sec. 26, WESTPORT, and the settlement around it. The name was formerly *Westport Station* but was changed by Dr. A. McDill to its present name, some time between 1873 and 1876. (Woodw.)

Mendota School

For *Mendota Station,* and ultimately L. Mendota. Between secs. 25 and 26, WESTPORT.

Mennes School ['mɛnɪz]

For Ole O. Mennes, Norwegian settler, and his descendants, who from before 1861 have owned the land on which it is: a former name for *Pleasant Hill School,* DUNKIRK.

*** Meora Creek**

This is one of the pseudo-Indian, or distorted names used only on Greeley's map (1854); it seems to apply to *Badfish Cr.,* but is not drawn in correctly. It had little if any currency.

Merrill Crest, subd. ['merəl]

For its location on high ground above *Merrill Park;* platted 1910, in NE sec. 19, MADISON.

Merrill Park, subd.

For Alfred Merrill, owner of the land; platted 1891, in SW sec. 17, MADISON.

Merrill Springs

For Alfred Merrill, who early owned adjacent land (see *Merrill Park*): springs in SE sec. 18, MADISON, so called from before 1861. Now within *Spring Park.*

Middle Branch (of Sugar R.)

For its position with respect to the upper part of Sugar R., and the *West Branch:* a frequent alternative name for *Deer Creek.* So called on Gree. map (1855), by Butt., and all platbooks since 1890 (Foote), which also labels another part *North Branch.* Foote and Gay (1899) also use "Middle Br." to apply to the *West Branch,* into which Deer Cr. runs.

Middleton, town

For the PO and village. Est. Mar. 11, 1848, by separation from MADISON; it includes T7N,R8E. (For the form of the name, see the PO.)

Middleton, vill.

For the PO. Since this has moved, the name has applied to two distinct settlements. The first of these was about the cross-roads or junction of roads, in secs. 22, 23, 26, and 27 (now MIDDLETON) first settled in 1841. Here *Middleton PO* was est. 1846. But when the PO was moved in 1870 to its present location (sec. 11), that settlement became Middleton. Thus the original village had to be renamed, and became successively *Middleton Junction* and *East Middleton.*

The second village had been successively *Peatville* and *Middleton Station,* but has been *Middleton* since the *Middleton PO* moved to it. It now includes parts of secs. 11 and 12. It was incorporated 1905.

Middleton Beach, subd.

For the town and nearby village of Middleton; on L. Mendota, platted 1909, in NE sec. 12, MIDDLETON.

Middleton Junction

For the junction of roads (and the presence of the *Junction House*): an early name for the original village of Middleton, the

"Junction" part (a frequent abbreviation thruout its history) perhaps antedating the PO and the "Middleton" part. When the PO moved and took the name "Middleton" to sec. 11, this became even more distinctively the Junction. See *Middleton PO.*

Middleton Junction School

For its location at the old *Middleton Junction.* Formerly *Red Brick School.*

Middleton PO

There is some uncertainty about this name, particularly the source of its present form, which has been taken by the village and town. In 1837 the "paper town" of *Middletown* was platted, and tho it came to nothing, it was not forgotten.

However, when the PO was est., 1846, it was given the name "Middleton." There is no recorded explanation of this change in spelling. It is said that the name was given by Harry Barnes, the first postmaster, in honor of the town in Rutland County, Vt.—but that is Middletown, so the change is still unexplained. It may be that the PO dept. objected, there being so many Middletowns already, and comparatively fewer Middletons (N. H., Mass., Mo., Ark.). Incidentally, there is no trace in early records of any person by either name in this locality. The cause of the change, then, is unknown. That there was uncertainty at the time is suggested by the fact that in the Laws of Wisconsin setting up the town in 1848, the law refers to "Middleton" but the marginal index spells it "-town." Lapham's map of 1848 also labels the town "Middletown." The PO, however, appears to have spelt it only "-ton."

Since the history of this PO and its competitors is very complicated, it is here given in detail. When *Middleton PO* was first est. (Sept. 22, 1846), it was kept at the crossroads on Mineral Point Road between secs. 22 and 27, MIDDLETON; it was disct. Aug. 8, 1856, but re-est. there Sept. 22, 1856. The first competitor appeared later that year (Dec. 10) in *Peatville PO,* sec. 11, where the present vill. of Middleton is. Two years later (May 13, 1858) a second competitor was est.: *West Middleton PO,* kept in the NE corner of sec. 30. (This was moved before 1890 to the crossroads half a mile E, between secs. 29 and 32, where it stayed until it was disct. Feb. 27, 1894; it was succeeded in the latter place by *Barwig PO.*)

Peatville PO meantime changed its name to *Middleton Station PO* (June 6, 1862), and this again changed the same year (July 30) to *Mendota PO*, to remain so for about 8 years, until (Apr. 26, 1870) it at last won the name "Middleton PO" away from the original location (secs. 22 and 27) which simultaneously changed to *Junction PO*. Thus Middleton PO achieved its present location. But Junction PO lasted less than 2 yrs.; on Jan. 17, 1872, it became *East Middleton PO*, to remain so till Dec. 22, 1893, when it was discontinued.

* **Middleton Station,** vill.

For the town and the station of the CMSP railroad, newly built: the village, platted 1856, which superseded *Peatville*, in sec. 11, MIDDLETON. It became the present *Middleton*. See also *Middleton PO*.

* **Middleton Station PO**

For the village and station. The name was changed to this from *Peatville PO*, June 6, 1862; this was changed to *Mendota PO* July 30, 1862. See *Middleton PO*.

* **Middletown**

Probably reminiscent (there being Middletowns already in Conn., Del., N. Y., Pa., Md., Va., Ohio, Ky., N. J., and N. H.): a "paper town," platted 1837, in NE sec. 9 and SE sec. 4 of the present MIDDLETON, by F. W. Heading, J. A. Noonan, and Sylvester W. Dunbar. The plat shows *Silver Lake* in the midst, and projected canals, roads, and railroads connecting it with everywhere. It never came to anything, but was probably partly the source of the name of the village and town of *Middleton*. Still referred to in 1845 (Dane 120).

Midland School

For its location close to the center of the township, in NW sec. 22, BLACK EARTH.

Midway, subd.

For its position along the highway and about midway between Madison and Middleton? Platted 1927, in SE sec. 17, MADISON.

Military Road, The (now also, **The Old Military Road**)

For its original purpose, to connect forts Howard (at Green Bay), Winnebago (at Portage), and Crawford (at Prairie du Chien). It passed thru Dane County from Blue Mounds to Mount Horeb, where it turned NE past Pine Bluff, thru Cross

Plains and BERRY, SPRINGFIELD, DANE, and VIENNA at Hundred Mile Grove. It was surveyed in 1832, built in 1835 (Smith). Many parts of the original course are now abandoned.

Mill Addition, The, subd.

For the nearby mill and Mill St.; platted 1923 by E. D. Hering, within the village of Cross Plains.

* **Mill Creek Prairie**

For Badger Mill Cr.; apparently an older name (1844) for *Badger Prairie.* (Dane 93, 97)

Miller Park, subd.

For C. L. Miller, owner of the land, who had it platted 1923; NW sec. 35, MADISON.

* **Millwood (or, Milwood)**

Descriptive: a settlement where Philo Dunning owned a sawmill, built 1841, in the woods along *Clyde Cr.*, in about sec. 33, BURKE. (Park, 423, 551; Dane 71)

* **Milwaukee and Baraboo Valley Railroad** [ˈbærəˌbu]

For the regions it was intended to serve: a former name for the *Watertown and Madison RR.* (Barton 1099)

* **Milwaukee and Mississippi Railroad**

For the intended termini: a former name (from about 1850) for the *CMSP RR.*

* **Milwaukee and Prairie du Chien Railroad**

For its termini: a former name of the *CMSP RR.*

* **Milwaukee and Western Railroad**

For the region it served; a former name for the *Watertown and Madison RR*, now a part of the CMSP. (Barton 794)

* **Milwaukee Madison Road, The**

For its termini: the first such road, also called the *Blue Mounds Rd.* (1)

Milwaukee Road, The

A current abbreviation for the CMSP railroad.

Mineral Point Road, The

Because it joined Madison with Mineral Point: an early road, part of which (from Pine Bluff to Blue Mounds) coincided with the *Military Rd.*, and part (from Pine Bluff to Madison) with the *Speedway Rd.* The name has been in use since the earliest days of Madison's settlement.

Minette's Hill [mɪˈnɛts]

For L. E. Minette, owner of part of the land since before 1911:

a large hill or bluff in sec. 30, ROXBURY, around whose S face the road curves.

Minniwakan Spring

From Dakota ("mi'-ni-wa-kaŋ . . . water spirit, i.e., whisky" Riggs): a spring in *Spring Park;* SE sec. 18, MADISON. The name is found only on Mendo. map, 1900 and Rural, 1910. Clearly not a native name, this must have been given by some white person; it may be wondered whether the namer was aware of the second meaning of the word.

Moen School ['moən]

For S. O. Moen, owner of the land (from before 1890 till after 1904) on which the school was built; SE sec. 31, BLUE MOUNDS.

Moline Park, subd. [ˌmo'lin]

For the Moline Plow Co., of Moline, Ill., which bought out the T. G. Mandt Wagon Co., a former manufacturing firm of Stoughton (Dow); platted 1902, within the city of Stoughton.

Monona, subd. (BLOOMING GROVE) [mə'nonə]

For nearby L. Monona; along the Yahara R., within the city of Madison, platted 1891; NW sec. 7.

Monona, subd. (MADISON)

For the lake, on which it borders; platted 1899, in NE sec. 26; since absorbed into the city of Madison. Since the plat was made by Geo. S. Lawrence, and the principal street is Lawrence St., he probably named the subdivision.

*** Monona Assembly**

For its being the site of yearly religious meetings (1881–1906): grounds along the shore of L. Monona, sec. 25, MADISON, formerly called *Lakeside* (now Olin Park).

Monona Bay

For L. Monona, of which it is the west end, distinctively surrounded by land on three sides, and crossed by two railroad tracks to the E; it covers parts of secs. 23 and 26, MADISON.

Monona Bay, subd.

For its position along *Monona Bay;* within the city of Madison, platted 1906.

Monona Crossing

The name used on the USGS map (1904) for *Black Bridge Crossing* (which it places erroneously).

Monona Heights, subd.
Descriptive: on high land along L. Monona; platted 1904, in SW sec. 17 and SE sec. 18, BLOOMING GROVE.

Monona Lake
= *Lake Monona.*

Monona Lake Assembly
= *Monona Assembly.*

Monona Lake Bay
= *Monona Bay.*

Monona Lake Ridge
So called by Chap., 1859. = *Monona Ridge.*

Monona Park, subd.
For its proximity to L. Monona; platted 1919, in SW sec. 25, MADISON.

Monona Ridge
For its location: a piece of high ground along the edge of L. Monona, in NE sec. 19, BLOOMING GROVE. See *Monona Lake Ridge.*

Monona Ridge, subd.
For *Monona Ridge*, on which it is; platted 1910.

Monona Village
For L. Monona, on which it borders: a village, incorporated 1938, including parts of secs. 8, 9, 17, 18, 19, 20, and 29, BLOOMING GROVE.

Monroe Park, subd.
Perhaps for Monroe, Green Co., with the idea of selling lots here to people from that village: cp. *Edgerton Beach;* platted 1894, in NW sec. 29, PLEASANT SPRINGS.

Monson's Park, subd. [ˈmʌnsn̩z]
For Martin P. Monson, owner of the land; platted 1923, in SW sec. 18, PLEASANT SPRINGS.

Montjoy (also, representing a local pronunciation, **Mountjoy, Mount Joy**) [ˌmɑntˈdʒɔɪ, ˌmɑʊntˈdʒɔɪ]
From the maiden name of the wife of the owner, Edward A. DeBower (DeBow.): a large stock farm in SW sec. 6, VIENNA, locally important from the first as the Abram A. Boyce property; owned by DeBower from before 1911 till after 1931; recently changed to *Dunrovan.* Highway maps show it as if it were a village.

Montrose, town [ˌmɑnt'roz]

For Montrose, Susquehanna Co., Pa. Accounts vary as to the man responsible for the choice: Durrie (404) names George McFadden; Park (479) names Peter W. Matts. The latter is more likely, since Matts was from Bucks Co., Pa.; it is also said that McFadden offered the name *Grand Springs*, which was refused by the legislature because too many other names already had the element "spring." When first est. Feb. 11, 1847, by separation from MADISON, Montrose included T5N,R6,7&8E; but when PRIMROSE and PERRY were separated, Mar. 21, 1849, it was reduced to its present area: T5N,R8E.

*** Montrose,** vill.

For the town: a "paper town" on the Sugar R., platted 1851, in adjoining parts of secs. 22 and 23, MONTROSE. It was never developed.

Montrose, vill.

For the PO: a settlement, mostly in sec. 30 (MONTROSE) but also in NW sec. 29, where modern maps still indicate it, and where Montrose School is. The PO was est. 1878, and the name must have been attached to the settlement at that time.

*** Montrose PO**

For the town. Est. July 18, 1878; disct. Aug. 31, 1900. Kept at the store in NE sec. 30, now on highway 92 (Fritz, Swig.).

Montrose School

For the former *Montrose* village, near whose site the school is: NW sec. 29, MONTROSE.

Morehead School

For Wm. M. Morehead, of Ohio, a prominent early settler, who took up from the government in 1845 the land on which the school stands: on the town road between secs. 22 and 23, MONTROSE.

Morland Terrace, subd. ['morlənd]

By back-formation from the earlier *Westmorland*, to the E of which this was platted, 1928, within the city of Madison; SW sec. 21, MADISON.

*** Mormon Baptismal Pond**

Descriptive: a spot on the upper part of the west branch of Sugar River, where it crosses from sec. 35, CROSS PLAINS to sec. 2, SPRINGDALE. It was used for their rites by some Mormons who settled near here about 1846–7. (Park 309, 444)

Morningside Heights, subd.
Descriptive (it is located eastward of L. Monona and the city of Madison) and probably also reminiscent; platted 1924, in secs. 9 and 16, BLOOMING GROVE.

Morrison, vill. ['mɔrɪsən, 'mɑrɪsn̩]
For James Morrison, from Scotland, first settler (June 6, 1843) on sec. 6, WINDSOR, in the NW quarter of which the village was platted, 1872. This is still the name of the plat, tho it is never used for the village; instead, the name of the PO (Morrisonville) has become the only name for the village.

Morrison Station
For its location in the village of Morrison: a station on the CMSP railroad, est. 1871. James Morrison I, gave 42 acres of land to the railroad company for platting the village if it would establish a station here. The station name was often used for the village.

Morrisonville, vill.
For the PO, est. 1871. This name has displaced the name of the plat, *Morrison*, in actual usage.

Morrisonville PO
For James Morrison, first postmaster, and founder of the village of *Morrison*, at which the PO was kept. Est. May 10, 1871. The PO was not called Morrison because there already was one near Green Bay.

Morris Park, subd. (DUNN) ['mɔrɪs, 'mɑrɪs]
For Morris Brown, an early local settler (before 1861) and former owner of the land; on L. Waubesa, platted 1899, in SE sec. 8 and SW sec. 9. Some platbooks spell Brown's name "Maurice," but the early records show that "Morris" was the correct form, and the plat uses this.

Morris Park, subd. (WESTPORT)
For W. A. P. Morris, former owner of the land; on L. Mendota, platted 1909, in SW sec. 33.

Morse Pond [mɔrs]
For J. W. Morse and his descendants, owners from before 1861 till after 1931: a pond in the SW quarter of sec. 3, VERONA. The name is shown on maps only of 1906–7 (USGS) and 1911 (Cant.), but was surely in use much earlier.

Mosquito Island
Because mosquitoes were plentiful here: an island in the Wisconsin R. in NW sec. 7, ROXBURY, just above the Sauk City

bridge. Now also called *Otto's Is.* and *Hiddesen's Is.* (Derl.)

*** Mound Fort**

= *Blue Mounds Fort.*

Mound Park, subd.

For the Indian mound located in it; along L. Monona, platted 1891, in SW sec. 9, BLOOMING GROVE.

*** Mound Prairie**

Descriptive; there are Indian mounds on it: a prairie covering parts of secs. 15, 16, 21 and 22, DUNN. Mentioned only by Lorin Miller, surveyor, on his map and in his field notes, 1833–4.

Mound Prairie

= *Nine Mound Prairie.* (Park, 572)

Mounds Creek

From its origin in the region of the *Blue Mounds:* it has an *East Branch* in Dane Co., and a West Branch in Iowa Co.; they join in the latter to flow into the Wisconsin R. near Arena. See *Birch R., Osakaw R., Pine R.*

*** Moundville,** vill.

For the *Blue Mounds*, SE of which it was, perhaps on sec. 8. This was apparently the first name applied to any settlement here, apart from "*Brigham's.*" It appears on a MS. map as early as 1828? (Farmer), on a printed map first in 1829 (Chan.), and on others as late as 1845 (Tanner); no plat has been found, however. The rival *Beaumont* was platted in sec. 6 in 1836. *Moundville PO* was est. 1837, and continued till 1839. The PO name was then changed (and presumably its location too) to Beaumont. Just how long the name "Moundville" continued in use is unknown.

*** Moundville PO**

For the village: the first official PO in BLUE MOUNDS; est. Jan. 28, 1837; changed to "Beaumont PO," May 18, 1839. See *Blue Mounds PO.*

*** Moundville Precinct**

For the village and PO: an election precinct est. May 16, 1839. (Dane) It was superseded by the *Blue Mound Precinct.*

Mount Horeb, vill. [ˌmaʊnt 'horɛb, maʊnt 'horəb]

For the PO. The settlement dates back to at least 1850; it centered at first about the crossroads near the middle of sec. 12, BLUE MOUNDS, but with the coming of the railroad was extended westward. Under this name, a part in SW sec. 12 was platted (1880) then, with *Mount Horeb Station* and other plats, formed an incorporated village (1899) which now includes all of sec. 12, and parts of secs. 11, 13 and 14. Former names were *Staangi* and *Horeb's Corners*.

Mount Horeb PO

For the Biblical mountain, not clearly distinguished from Mt. Sinai, in the Sinai peninsula, Egypt: so named by the Rev. George Wright, the first postmaster. On the basis of *I Kings*, 19:8, and other texts, it was thought to mean "the Mount of God," tho actually the word means "desolate" in Hebrew. The PO, est. July 25, 1861, was first kept in SE sec. 1; later, at various points within the present limits of Mount Horeb village.

*** Mount Horeb Station,** subd.

For the village and the new RR station: 2 plats continuous with that called *Mount Horeb;* made in 1881, in SW sec. 12 and SE sec. 11, BLUE MOUNDS. Absorbed into the village when it was incorporated 1899.

Mountjoy

= *Montjoy.*

Mount Julia

Origin unknown; it is for a woman, but only the vaguest legends remain about her: a small isolated ridge rising about 200 ft. above the surrounding land; N sec. 24, PRIMROSE. The name appears on only one map (H&W); it is still well known, tho hardly used.

Mount Pleasant

Subjectively descriptive: an alternate name, now not much used, for *Helm's Hill.*

Mount Pleasant School

Subjectively descriptive. At the crossroad where secs. 25 and 36, DUNKIRK, and 30 and 31, ALBION, meet.

Mount Vernon, vill.

For George Washington's home, in Virginia, the native state of Joel Britts, and George, his nephew, who were the people most instrumental in starting the village. Joel Britts came in 1848 and took up the land; the village was platted 1852. In 1852 George also built a sawmill nearby, and later a flour mill in *Britts Valley* (Gilb.). The plat included parts of SW and W sec. 34, SPRINGDALE, and NW sec. 3, PRIMROSE.

Mount Vernon Bluff(s)

For the village of Mt. Vernon, just across the creek: the W

bluffs of Deer Creek, chiefly in sec. 33, SPRINGDALE. Also called *Mt. Vernon Rocks*.

Mt. Vernon Creek

For the village thru which it flows: a local name for what maps show as *Deer Cr.* (Austin)

*** Mount Vernon PO**

For the village. Est. June 14, 1854; disct. Aug. 30, 1902.

Mount Vernon Rocks

= *Mount Vernon Bluff(s)*.

Mount Vernon School

For the village; the school is in SE sec. 34, SPRINGDALE.

Mrs. Hansen's Park [ˌmɪsəz ˈhænsn̩z]

For Mrs. A. Hansen, whose land adjoins it, and who has been there since before 1911: the part of *Blue Bill Park* which is on the E side of the Yahara R. So known locally. (Corc.)

Mud Creek

Descriptive: a small creek rising in the N part of CHRISTIANA, flowing N thru DEERFIELD, and joining Koshkonong Cr. on sec. 23. So named before 1880, probably much earlier. (Butt.)

*** Muddy River**

Descriptive: a former name of *Sugar River;* only on maps of Farmer (1830) and Mitch. (1835).

Mud Lake (ALBION)

Descriptive, with a touch of humor: a small lake or pond in SW sec. 34. Shown without name on many early maps from 1835 (Terr.) forward; the name first shown in 1890 (Foote).

*** Mud Lake** (BRISTOL)

Descriptive: another name for *Norwegian L.* or *Bornson L.* (Pfaff).

*** Mud Lake** (COTTAGE GROVE)

Descriptive: a small "spread" in a creek tributary to Koshkonong Cr., in E sec. 15. So named on platbooks of 1911 and 1926; it is now dried up.

Mud Lake (DEERFIELD)

Descriptive: a small lake, or pond, in NW sec. 2. First shown on map of 1861 (Lig.) without name; first with name, 1873 (H&W).

Mud Lake (DUNN)

Descriptive: a "spread" in the Yahara R. Between lakes Waubesa and Kegonsa; secs. 10 and 11. Shown without name on early maps from 1830 (Farmer) forward; first with name 1861 (Lig.).

* **Mud Lake** (MIDDLETON)

Descriptive: a shallow body of water, the source of Black Earth Cr., as late as 1900 covering a considerable tract, chiefly in secs. 8 and 9. The name was in use as early as 1861 (Lig.) and as late as 1916; the swamp has been drained since. See *Silver Lake*.

Mud Lake (ROXBURY)

Descriptive: properly, that arm of *Fish Lake* which stretches W of the main part into N sec. 4; it is now connected with the main part only by a culvert under the road. (Mack, Mussen)

* **Mud Lake School**

For the nearby *Mud Lake:* a former name, current in 1911, for *Twin Valley School*, MIDDLETON.

Muir Knoll [mɪur]

For John Muir, "naturalist and father of the national park system" (Barton): a high point on University Hill, on the University of Wisconsin campus; sec. 14, MADISON. The name was suggested by Prof. Julius Olson when, in 1918, a boulder was placed to mark the Knoll, across the drive from North Hall, in which Muir lived as a distinguished student at the University of Wisconsin, 1860–4.

* **Munsel Spring** [ˈmʌnˌsel]

For a family by this name, which owned the land before 1861: a spring in SE sec. 14, MONTROSE. (Boning)

Murphy School

For Edward Murphy, Irish settler, and his family, who have owned land hereabouts since 1855. The school is in W sec. 29, CROSS PLAINS.

Murphy's Creek (FITCHBURG, DUNN)

For A. Murphy; see *Murphy's Spring:* a small creek rising on sec. 22 (USGS), FITCHBURG, and flowing E, then NE thru DUNN into L. Waubesa. Formerly also *Penora Cr.*

Murphy's Creek (also, **Murphy——**) (MADISON)

A more recent alternate name for *Wingra Cr.*, particularly the lower part. The name appears on a map first in 1904 (USGS, "Murphy Cr."); but it may be connected with J. Murphy, who owned land before 1873 in the nearby NW quarter of sec. 36 (H & W).

Murphy('s) Spring

For Abraham Murphy, an early settler here (1844), and his descendants; a spring in NE sec. 24, FITCHBURG. Its water flows into *Murphy's Cr.*

* **Muskrat Creek**

For the presence of many muskrats in the marsh about it: the earliest name of *Saunders Cr.* Lorin Miller, first surveyor (1833) of the present ALBION, writes in his field notebook that this creek was "called Muskrat Cr."—so the name antedates the survey, and was probably given by some trapper. It is mentioned again only by Butt., but probably as a historical name.

* **Myer's Corners** (also, **Meyer's** ——) [maɪɚz]

For Hartwig Myer, or Meyer, storekeeper at the site of the present Marxville from before 1861. This name was the general one for the settlement, from before 1877, before the establishment of Marxville PO there (1886), and probably continued in use for some time after.

Nakoma, subd. [nə'komə]

Chosen by Alfred T. Rogers, chief promoter, from Baraga's dictionary of Ojibwa, in which it is given as meaning, "I promise him to do something." For promotional purposes this was changed to "I-do-as-I-promise," and made the "official" meaning (see the *Nakoma Tomahawk*). First platted 1915; now includes part of secs. 28, 29, 32 and 33, MADISON, and is part of the city of Madison.

* **Natural Mounds**

Descriptive: two prominent hills in the NE corner of DUNKIRK. Tho shown on several early maps, 1835 Terr. is the only one to label them; thus this may not be a name at all, but only a descriptive designation. They have no names today, tho recognized as among the highest points in the town.

* **Nauneesha Ford** ['nɔnɪˌʃɔ]

A crossing of the *Nauneesha River*, NE sec. 25, BRISTOL. The name is used only by Eldred, and seems a conscious archaism.

* **Nauneesha River** (also, **Nauneeshah, Naunesha**)

Origin uncertain; possibly from Ojibwa *nānīcā* [nɔ'niʃɔ], meaning "divided several times", i.e., the river having parallel channels separated by islands (Geary, Lincn.). The alternate form *Mauneeshah* is a mistake; it does not appear till 1847, whereas the forms beginning with N are on maps of 1839 (Cram), 1844

(Doty), and 1846 (Lapham), and in Doty's letter of 1840. The M- forms, however, have to some extent survived, whereas the N- forms are virtually unknown. The present official name is *Waterloo Cr.*

*** Neosho Creek**

Of uncertain origin; perhaps for the previously named Neosho River, a tributary of the Arkansas, which has given its name to many places in Kansas and Missouri. The name appears only on Greeley's map of 1854, on which there were many other nonce names of doubtful origin. Brown gives the word as Winnebago ([nihoʃu] means "containing water," confirmed by Stucki); there is also said to have been a Winnebago village of the same name on the Yahara R. (into which this flows) and the adjoining N shore of L. Mendota (Brown 14). This name gained no currency; the present name is *Six Mile Creek.*

Nessa School

For O. Knudson Nessa and his descendants, who owned adjoining land from before 1861 till after 1904. The school is in NW sec. 34, PRIMROSE.

New Halfway Prairie School

For the prairie and creek. In NE sec. 8, BERRY. It was built more recently than *Old Halfway Prairie School.*

*** New Janesville Road, The**

Because it was the later road between Madison and Janesville; see the old *Janesville Rd.* This was built in 1845 (Dane 127) and is commonly known today as the *Oregon Rd.*

Nilsedalen Valley ['nɪlsəˌdɑ˖lɛn]

Norwegian for "Nils's valley"—i.e., for Nils Raanum (on plat books spelt Raunum, Ronum, etc.), who settled there (NE sec. 16, Perry) in the 1890's. (Haugen)

Nine Mound Prairie

For nine circular Indian mounds discovered thereon, and named, in the summer of 1840, by a party of 10 or 12 of the early settlers of VERONA, who had set out to explore the upper Sugar R. valley. The mounds, now nearly obliterated, are in SE sec. 8 (there is a tenth in the form of an animal), and the prairie stretched thru most of secs. 5, 6, 7, 8, 17, 18, 19, 20, 29, and 33. The name first appears in Lig. (1861).

Nine Spring Marsh

For Nine Springs Creek, which flows thru it: NE FITCHBURG and SW BLOOMING GROVE.

Nine-Spring(s) Hill (also, **Nine-Spring Marsh Hill**)

For Nine Spring Creek and marsh, at its foot: a sharp hill over which the Madison-Oregon Rd. passes; mostly in S sec. 1, FITCHBURG (Barton 1078–9). A later alternate name for *Break Neck Hill.* See also *Sutherland Hill.*

*** Nine Spring Prairie**

For its location a short distance S of the creek: an alternate name for *Syene Prairie* (Butt. 1257).

Nine Springs, The

Descriptive: the head of Nine Springs Creek. Actually, there are many more than nine springs, tho perhaps nine main ones. (Secs. 3 and 10, chiefly; FITCHBURG.)

Nine Spring(s) Creek

For the *Nine Springs* which form its source: a creek rising in the N part of FITCHBURG, and flowing E into the Yahara R. between lakes Monona and Waubesa. First recorded as "Nine Spring Run" (1844: Dane 82), then as "Nine Spring Branch" (1844 and after: Dane 97 ff.); Lig. (1861) makes it "Nine Spring Creek." See *Tarpora Cr.*

Nissedahle [ˈnisəˌdɑ˖lə]

For Nissedal (Telemark, Norway), with the ending altered to conform to the surname of Isak Dahle, its founder and namer, whose family was from Nissedal. In 1927, shortly after his return from a trip to Norway, Dahle bought the land (on NW sec. 4, BLUE MOUNDS) and set out to make there a reproduction of a typical Norwegian farm. Dahle gives as the meaning of the name "the Valley of the Nymphs," but this is a recent folk-etymology; "Nissedal" is from Old Norse "Nizidalr: the valley of lake Nizir" (Haugen). "Nissedahle" is in limited local use; the common name is *Little Norway.*

Nonn's Corners [nʌnz]

For Jacob Nonn, owner of the land since some time between 1899 and 1904: a crossroad in SW sec. 19, SPRINGFIELD. Formerly *Lappley's Corners*, and before that, *Schafer's Corners.* (Dahmen)

Nonn School

For Jacob Nonn, and kept at *Nonn's Corners.*

Nora Corners

For *Nora PO*, which was moved here in 1886: a crossroads settlement between secs. 35 and 36, COTTAGE GROVE. (The name may have been used as a settlement name, as well as a PO name, before the PO was moved into COTTAGE GROVE; see Park 439.)

*** Nora PO** (also, erroneously, **Norah**)

A name intended to suggest Norway. When A. A. Prescott, the first postmaster, was granted a PO, he, a Norwegian in a very Norwegian settlement, wanted to give it a Norwegian name. Those suggested at first seemed too un-English, but someone, beginning with the first syllable of Norway and Norwegian, hit upon this, which was approved (Reque). The PO was est. Apr. 13, 1869, and was at first kept on Prescott's land in SW sec. 29, DEERFIELD. In 1879 it was moved to where *Sweet Home PO* had been. In 1886 (Thor.) it was again moved, to the present *Nora Corners* (COTTAGE GROVE). It was disct. May 19, 1893; re-est Oct. 26, 1893; disct. Aug. 3, 1902.

Nora School

For *Nora Corners*, at which it is.

Nordness Corner ['nɔrnɛs] (Juve)

For M. L. Nordness and his descendants, who have owned adjoining land since before 1890: the road junction between secs. 21 and 28, PLEASANT SPRINGS.

North Bay, subd.

For its location near a bay of the NE shore of L. Mendota, in sec. 12, MADISON; platted 1927; it later became a part of the village of *Maple Bluff*.

*** North Branch (of Sugar River)**

For its position with respect to the other branches: a former name of the upper part of *Sugar R.;* so called by Butt. (1880) and Foote (1890). Foote also labels *Deer Cr.* with this name.

North Bristol, vill.

For its location around the cross roads between secs. 5 and 8 BRISTOL. The name did not become official till the establishment of the PO (1876), but settlement was made many years before.

*** North Bristol PO**

For the village; est. Jan. 13, 1876; disct. July 31, 1910.

*** North Clinton**

See *North part of Clinton, The.*

Northfield Park, subd.

Partly descriptive (since the plat was made from a field) but also reminiscent of Northrup Stadium in Minneapolis, in which city C. B. Chapman, who gave the name, had lived for some years (Frey). Platted 1940, in NW sec. 6, BLOOMING GROVE.

North Gardens, subd.

For its position within the city of Madison; perhaps also influenced by the adjoining *North Lawn;* platted 1928.

North Lawn, subd.

For its position within the city of Madison, and probably also to correspond with the earlier *West Lawn;* platted 1916.

*** North Madison**

For its position with respect to Madison: a "paper town," shown by Judson (1836) about the NE part of L. Mendota, and including secs. 1 and 12, of the present MADISON, and secs. 34, 35 and 36 of WESTPORT. No plat was recorded, and it was never settled as such.

North Park, subd.

For its location along the N shore of L. Mendota; platted 1907 in N secs. 35 and 36, WESTPORT.

*** North part of Clinton, The**

For its location with respect to *Clinton:* a "paper" addition to a village which itself remained in the "paper" stage for 10 years. Platted Dec. 12, 1836, as "The North part of Clinton," tho also referred to as "North Clinton," it was to be in SW sec. 13; CHRISTIANA.

North Shore Bay

For its location on the bay in the N shore of L. Mendota: the name, since about 1938, for *Second Ward Beach.*

North Side, subd.

For its position within the city of Madison, this being in the NE corner of the original plat; platted 1897; SE sec. 12 and NE sec. 13, MADISON. The term "North Side" may have been applied to this part of the city before this particular plat was made.

*** Northumberland Settlement**

For Northumberland, Saratoga Co., N. Y.? There are no records of English settlement here, but one of the earliest settlers Jacob Ostrander, who took land on sec. 17, BRISTOL, 1844, was a native of Rensselaer Co., near Saratoga. This settlement is mentioned only by Park (317); it is unknown today. The settlement became a Norwegian one, and there may be some allusion to that intended in the name.

North Windsor

For its location: generally the N, particularly the NE part of WINDSOR. Adjoining secs. 2 and 11, in which are the North Windsor Methodist Church and the North Windsor School, are about the center of the area.

* **North Windsor,** vill.

For the PO, est. 1865 at this settlement in NW sec. 17. The village was platted in 1874 as *DeForest.*

* **North Windsor PO**

For its location in the town. Est. May 18, 1865, it was kept in NW sec. 17 on the land of Isaac N. DeForest, the first postmaster. It was changed to *DeForest PO* Dec. 3, 1872.

North Windsor School

For its location in *North Windsor:* extreme SE sec. 2.

Norway Grove

For an early settlement of Norwegians which began there in 1847: a grove which originally covered more than one fourth of VIENNA (all of secs. 21 and 22, and parts of adjoining 14, 15, 16, 23, 26, 27, 28, 33, and 34). Much of this has since been cleared as farmland.

Norway Grove, cor.

For the grove: a small crossroads settlement in adjoining N parts of secs. 22 and 23, VIENNA.

Norway Grove Church

For the *Norway Grove* settlement: the "Norway Grove First Evangelical Lutheran Church of De Forest"; est. before 1861, in SE sec. 24, VIENNA.

* **Norway Grove PO**

For the grove and settlement. Est. Mar. 21, 1872, and kept in the extreme NE corner of sec. 22, VIENNA; disct. June 30, 1906.

* **Norwegian Hill**

For the origin of its settlers: a hill mostly in N and W sec. 15, BURKE, crowned by a Norwegian church on the section-line road which runs over its W slope. The name is mentioned only by Park (426).

* **Norwegian Lake**

Because Norwegians owned the land around it: a name among the German settlers and others for the small lake (now drained) in sec. 18, BRISTOL, also sometimes called *Bornson Lake.* (Pfaff)

Oak Beach, subd.

Descriptive; on L. Mendota; platted 1910, in E sec. 12, MIDDLETON.

Oak Crest, subd.

Descriptive; platted 1945, in N sec. 19, MADISON.

Oak Edge School

For its location on the edge of oak woods; in NW sec. 5, BURKE; now closed.

Oak Grove, subd.

Descriptive; platted 1922; NW sec. 25, MADISON.

* **Oak Grove,** cor.

For *Oak Park*, cor.; this name is used only on the USGS map, 1905.

Oak Grove School

For a nearby group of oak trees: in NW sec. 36, MONTROSE.

Oak Hall (Corners), (also, **Oakhall**)

From the name of a hotel built (about 1860 or before) at the NE corner of the crossroads by a man named Wood (Barton 1078): the settlement at the center of sec. 33, FITCHBURG; formerly *Fitchburg Corners.*

Some maps show an Oak Hall PO here, but there never was such. *Fitchburg PO* was first kept just S of here, and *Floyd PO* here. See *Floyd*, cor.

* **Oak Hall PO**

No such PO was ever est.; this is an error on some maps, which follow the local custom of so speaking of *Floyd PO,* kept at Oak Hall Corners.

Oakhall School

For *Oak Hall* corners, at which it is.

Oak Hill School

Descriptive; it is on a slope having oaks. At the road junction where secs. 7, 8, 17, and 18 meet, RUTLAND. Now closed. The older name is *Tipple School.*

Oakland Heights, subd.

Descriptive; platted 1896, in S sec. 22, MADISON; since absorbed into the city of Madison.

Oak Lawn School (DUNKIRK)

Descriptive. In SE sec. 30. Formerly *Frog Pond School.*

Oaklawn School (also, **Oak Lawn** ——) (SUN PRAIRIE)

Descriptive: it has a lawn with oak trees. In SW sec. 2. Formerly *Peckham School,* and *Sweet School.*

Oak Lawn School (YORK)

Because there are oaks and a lawn. In SW sec. 9. Now closed.

Oak Park

Descriptive, and probably also reminiscent: a resort on L. Monona, never platted, but shown on platbooks from 1899 forward; SW sec. 17, BLOOMING GROVE.

Oak Park, cor.

Descriptive: a crossroad settlement in NW sec. 17, DEERFIELD. In the 1890's a creamery was built here, which became a tavern, store, etc. *Adsit PO* was moved south to be nearer to this settlement. (Thor.) See *Oak Grove,* cor.

Oak Ridge, subd.

Descriptive; platted 1908, in N sec. 35, MADISON.

Oakridge, subd.

Descriptive: an addition to Mt. Horeb; platted 1914, NE sec. 11, BLUE MOUNDS.

Oakside School

Descriptive: there are oak trees nearby. In SE sec. 20, DUNN. Formerly *Keenan School.*

Oak View School

For the surrounding oaks; SW sec. 18, BRISTOL.

Observatory Hill

From its being the site of the astronomical observatory of the University of Wisconsin (built 1878): a prominent hilltop on the University campus; sec. 15, MADISON.

* **Odell Corners** [o'dɛl]

For George Odell, who bought the land formerly owned by Wm. Mayhew: a later name for *Mayhew's.* Odell and his family were here from before 1861 till after 1873.

Odland School ['ɑdla⊥nd]

For Ole K. Odland, local farmer, who was one of the first children of Norwegian parentage in this settlement; he was born in 1847, in NE sec. 1, DUNKIRK. The school is in the same quarter-section. Odland's father was named Knudson, and Odland himself did not possess the land till some time between 1873 and 1890; so the school name probably does not antedate this, tho the school was built before 1873.

*** Ohio Settlement, The**

For the many members who had come from Ohio, prominent among whom were Miller and Elan Blachly, A. J. Luce (or Luse), Wm. Dunlap, and Samuel Bell: a settlement in the W part of DANE, begun 1843.

Olbrich Park Addition ['ol,brık]

For Olbrich Park, which it adjoined; ultimately for Michael Olbrich, a prominent Madison attorney; platted 1940, in NE sec. 8 and NW sec. 9, BLOOMING GROVE.

Olday's Hill (also, **Oldy's——**) ['oldız]

For Frederick Oldy (later spelt "Olday") and his descendants, owners of some of the land since the 1890's: a sharp hill over which the road runs between secs. 22 and 27, VERONA. (Davids.)

Old Cottage Grove

For the original settlement named *Cottage Grove;* now *Vilas.*

Old Deerfield, cor.

As the site of the original village of *Deerfield.*

Old Deerfield School

For nearby *Old Deerfield.* In SE sec. 9, DEERFIELD.

Old Halfway Prairie School

Because it was the older, though it can hardly have been so called before *New Halfway Prairie School* was built. It was formerly Halfway Prairie School. SE sec. 11, MAZOMANIE.

*** Old Indian Garden**

For the gardens which the Winnebagoes had here until the whites came. A jut of land into the E part of L. Monona: secs. 19 and 20, BLOOMING GROVE (Butt. 927). See *Wood's Point, Winnequah, Belle Isle.*

Old Lake, The

Because it is now less large and less lake-like than it once was: this is probably the most current local name for *Brazee's L.*

Old Pompey

= *Pompey Hill.* (W. and F. Kenn.)

Old Roundtop

= *Roundtop* (DANE)

*** Old Smoky**

Descriptive (see *Smoky Mountains*): an alternate early name for the Blue Mounds, said to have been used by the Indians (Barton 1026–7), but cp. the two preceding names, and Appendix I, p. 202.

*** Old Town, The,** vill.

Because the first settlement of Mt. Horeb began here, and later settlement followed the establishment of the RR station farther W: the junction in the center of sec. 12, BLUE MOUNDS. (Barton 1084)

Ole Quam's Park

= *Quam's Park.*

Olson's Hill

For Tollef Olson (Norwegian settler who bought the land before 1873) and his family, owners of S sec. 15, then recently, 14, DUNN. This is an alternate name for *Daly's Hill,* and the more current one today (Coll.).

One Hundred Mile Grove

= *Hundred Mile Grove.*

One Hundred Mile Tree

= *Hundred Mile Tree.*

Orchard Lawn, subd.

Descriptive? Platted 1906, in NE sec. 26, MADISON; since absorbed into the city of Madison.

Oregon, town

For Oregon, then not yet a territory, but being rapidly settled. Separately organized out of ROME, Feb. 11, 1847: T5N,R9E. The petition for a separate town was circulated by Rosel Babbitt (Park 511); so perhaps it was he who suggested the name. It is interesting that the name Oregon was, according to the most acceptable explanation, a corruption of an early spelling of Wisconsin (Stewart), whence having traveled to the Pacific coast, it came back to its home state in the name of this town and its chief village.

Oregon, vill.

For the town: a village platted 1847 in NW sec. 12, OREGON; since enlarged to include parts of secs. 1 and 11 also; incorp. 1883. Formerly *Rome Corners.*

Oregon Branch (**of Badfish Cr.**)

Because it rises in the village of Oregon, OREGON. It runs E then S in RUTLAND, to join *Badfish Cr.* on sec. 16. So called from at least 1890, probably earlier.

Oregon PO

For the town and village. Est. Oct. 4, 1850.

Oregon Road, The

Because it joins Madison to Oregon: the present US highway 14 out of Madison to the S. Formerly *The New Janesville Rd.*

Orvold's Park, subd. (also, **Orvold** ——) ['ɔrvəld]

For Nels O. Orvold, owner of the land; along the Yahara R. and L. Kegonsa, platted 1907; SE sec. 14, DUNN.

*** Oskaw River**

Of uncertain origin; possibly from Ojibwa [oskɔ], or Potawatomi or Menomini [oskɔw], meaning "it is new," i.e., suggesting that the river has made itself a new channel (Lincn., Geary). It appears only on Center's map (1832), "Oskaw or Pine R.," where it is placed on the present *Mounds Cr.*—probably erroneously: see *Pine R.*

Otto's Island

For Mr. Otto, a recent owner: another name for *Mosquito Is.* (Derl.)

Pancake Valley

A well-known local humorous name for the *Vermont Valley.* Said to have been given in early days by a tramp who, having asked for food at several settlers' houses, was everywhere offered pancakes. (Ander., Pauls.) This story is apocryphal, but no other explanation is given. The valley is not strikingly flat.

Paoli, vill. [ˌpe'olaɪ]

For Paoli, Chester Co., Pa. The name was given by Peter W. Matts, born in Bucks Co., Pa., who laid out the village in 1849, on parts of SE sec. 3 and NE sec. 10, MONTROSE.

*** Paoli PO**

For the village. The name was changed from *Grand Spring PO*, Nov. 27, 1858; disct. Mar. 30, 1907.

*** Paoli Station,** vill.

For the nearby village of Paoli: a village, platted 1888; later changed to *Basco Station.* When the station was to be established here, the railroad officials wanted to call it "Boning," in honor of Henry Boning, a prominent early nearby settler, but he refused. (Boning)

Park Addition, subd. (DUNKIRK)

For the nearby Harrington Park; platted 1903 and added to the city of Stoughton.

Park Addition, subd. (BLUE MOUNDS)

For its bordering on Boeck Park (a part of the plat); platted 1905, in the village of Mt. Horeb.

Parkside, subd. (MADISON)

For its position along Tenney Park; platted 1905, within the city of Madison.

Parkside, subd. (DUNKIRK)

For the nearby Harrington Park; an addition to the city of Stoughton; platted 1912.

Park View, subd.

Because formed out of Schuetzen Park and adjoining Lake Park; within the city of Madison, platted 1902 by the Park View Land Co.

Paton School [ˈpætən, pætn̩]

For Adam Paton and his family, who from 1847 have owned land in this neighborhood. In NE sec. 9, VIENNA. Now closed.

Patrick's Lake (also, **L. Patrick**)

For Wm. W. Patrick, who settled on its W shore soon after 1840: a contemporary alternate name for *Brazee's L.* Patrick and Brazee were brothers-in-law, and there seems to have been a contest for the honor of giving the name. Brazee was more successful, getting his name used on maps and deeds; "Patrick's L." is not found on any map, tho it was once in wider use than "Brazee's L."; neither is much used today. See also *Duscheck's L., L. Washington, The Old Lake.*

Patterson Hill

For Jacob M. Patterson, of Ohio, and his family, who owned the land from before 1890 till after 1911: a hill in secs. 33 and 34, COTTAGE GROVE.

* **Peatville,** vill.

For the peat-beds in the vicinity (see Lig.). The name was probably given by Burgess C. Slaughter, who had an interest in digging the peat, for which kilns were set up, etc. Slaughter also became the first postmaster of Peatville PO, which was est. in the same year as the coming of the railroad and the platting of the village: 1856. The name was changed with that of the PO to *Middleton Station* (1862), but undoubtedly lasted in usage after that time. Whether the use of the name preceded the platting is unknown.

*** Peatville PO**

For the village. Est. Dec. 10, 1856; changed to *Middleton Station PO*, June 6, 1862. See *Middleton PO*.

Pecatonica River [ˌpɛkəˈtɑnɪkə, ˌpɪkəˈtɑnɪk] (variant forms below.)

There have been three fairly distinct forms of this name, as follows:

A. Pekeetennoe (also, Pee-kee-tau-no, Peektano, Pectanon); used between 1814 and 1835.

B1. Pick-a-tola-ky (also, Pocatolaka, Pee-kee-tol-a-ka, Pekeetolikee, Peckatolic, Pekatolika, etc.); used between 1828 and 1836.

B2. Peek a tón oke (also, Pekatonika, Peekatonokee, Peecatonnica, Pekatonic, Pecatonica, etc.); used between 1829 and the present, the present form having appeared first in 1856.

It will be noticed that B1 and B2 are no more than phonetic variants, the first having *l* in the stressed syllable, where the second has *n*. Thus the B forms, together, compose a single group distinct from that of the A forms. It would seem that two names of different structure and meaning but of similar sound were confused or mistakenly identified by the whites who got these names from the Indians—who themselves apparently spoke different dialects.

Judging on linguistic grounds, the *l* variants must come from Peoria (= Illinois) or Miami, and the *n* variants from Sauk, Fox, Kickapu, Potawatomi, Menomini, Ojibwa, or Ottawa. The word may be analyzed: *pīkw-*, many, cluster, group; *-at-*, shore, edge; *-ōlak-/-ōnak-*, canoe, boat; *-a-*, there is/are; *-wi* final third singular, which is dropped in all but Fox. (Geary)

Combining geographic with linguistic grounds, then, the B1 forms look like Illinois; the B2 might be Sauk, Kickapu, Ojibwa, or (much less likely) Potawatomi or Menomini, and the name means "many canoes grouped along the shore."

As to the A forms, it should first be noted that in 1673 Father Marquette recorded "Pekitanoui," meaning "Muddy River," as a name given to the Missouri. In present-day Fox the term *Pīkihtanwi* is still so applied, but since it is simply descriptive, it may be equally well applied to any muddy stream. Among the A forms is also "Pectanon," which appears to be the inanimate third plural meaning "muddy rivers," i.e., "rivers are muddy."

This may represent Fox *pīkehtanōni* or Ojibwa *pīktanōn*. (Geary) In any case, the A forms seem to be distinct in etymology and meaning from the B forms, and it is probably the coincidental similarity of the A to the B2 forms that led to the use of all three as alternates, implying that they were taken as no more than variants of the same basic form.

The fact that B2 forms have similarities with both the others, tho the latter are less alike, may well account for the survival of the B2 forms. The variety of alternants is found before 1836; by about that time the standard form had begun to rise. A forms disappeared after 1835, B1 forms after 1836, and B2 forms, preserving features common to both, became the accepted standard. Variant spellings subsisted until recent years, but custom has made "Pecatonica" prevail. At least two variations in pronunciation still exist, however.

The stream now called the Pecatonica has many branches and sources, some of the eastern ones of which flow from Dane Co. The name was applied rather irregularly at first, however, since its course was not known. In the treaty of 1829 "Pee-kee-tau-no" and "Pee-kee-tol-a-ka" are given as alternatives, and refer to the main stream of today, yet the former is used of the Sugar R. in 1835. Thereafter, Sugar and Pecatonica rivers are clearly distinguished.

This has been almost always called a "river"; it is called a "creek" only in the Treaty of 1828.

*** Peckham School**

See *Peckham's Corners*, half a mile north of which this school is. A former name for *Oaklawn School*.

Peckham's Corners ['pɛkəmz]

For J. A. Peckham and his family, whose land adjoined this crossroad (where secs. 2, 3, 10, and 11 meet, SUN PRAIRIE). They settled about 1845 and remained till after 1911.

*** Pectanon River**

See *Pecatonica R.* The identity of this stream was never clear, since such river names were usually applied before all the branches of a stream were known. The Treaty of 1829 used it as an alternate of "Pee-kee-tol-a-ka" (now Pecatonica), but Mitch. 1835 identified it with the *Sugar R.*, a branch of the Pecatonica. (The name "Sugar R.," however, antedates the use of "Pectanon" for that stream; in other words, there was overlapping usage for a time.) See also *Blue Mounds Branch.*

The latest use of this name on a map seems to have been 1835 (Mitch.), but long before this "Pecatonica" had come into being, and as this became established, "Pectanon" was discarded.

Peculiar, cor.

For *Peculiar PO.* The name is preserved on maps only, after the discontinuation of the PO.

*** Peculiar PO**

Some details of the story of the naming of this PO differ in various versions, but substantially the facts seem to be as follows. The PO was awarded to Thomas A. Denney, who wished to name it Denney PO, but vigorous local opposition arose, and for some time a decision could not be reached. Then when some names were submitted, they turned out to be the same as, or too much like, others already in existence. At last, the PO Dept. lost patience and wrote that it was peculiar, or that they were peculiar people, if they could not decide on a name; whereupon Denney suggested that they adopt the name "Peculiar," and so it was done. The PO was est. Sept. 26, 1898, and was kept at the road junction in SW sec. 5, VERMONT; disct. June 29, 1901. See Betty Cass in WSJ, 11.14.35, sec. 2, p. 1.

*** Peena Creek** (3 syllables: on the map as "Pe e na")

Said to be Winnebago for "Good Water Stream" (Brown); [pĩninɑ] = "good water" (Stucki). For *Pheasant Branch Cr.* This is one of the Indian names found only on Greeley's map of 1854. It gained little if any currency, tho Draper (who took part in choosing some of the names used first on this map) uses it in a footnote in 1856 (WHS *Coll.* 2:338).

*** Peirceville PO**

See *Pierceville PO.*

*** Penitto Creek**

Origin unknown. This is one of the distorted Indian names found only on Greeley's map of 1854, where it is applied to a creek shown as rising on sec. 21, BLOOMING GROVE, and flowing S into L. Waubesa. Early platbooks also show it (without name), but not the USGS map (Madison Quadr. 1904). The name gained little if any currency.

*** Penora Creek**

Origin uncertain: may be Winnebago; [pĩnora] = "the good body"

(Stucki). This is one of the distorted Indian names used first on Greeley's map of 1854; it may have had some currency: **Durrie** uses it, and, following him, Butt. Other maps leave the creek unnamed, but it is known locally as *Murphy's Cr.* (Barry)

Perry, town

For Commodore Oliver H. Perry, hero of the battle of Lake Erie, 1813. The name was actually proposed for the present PRIMROSE (T5N,R7E) by Ohioans living there, but the legislature, to avoid confusion, kept "Primrose" for the town having that PO, and gave "Perry" to the town next W (T5N,R6E) (Barton). Thus Primrose and Perry were separately organized out of Montrose, Mar. 21, 1849, but the latter remained attached to the former for legislative purposes till Nov. 24, 1853, when Perry achieved fully separate status.

Perry Log Church

For its location and construction: a Norwegian Lutheran church in NE sec. 7, PERRY; built 1851 and used continuously till 1887, it is now kept as a monument to the Norwegian settlers. (Ruste)

*** Perry PO**

For the town. Est. July 10, 1857, it was kept at first by the postmaster, Anaun Sanderson, on his land in NE sec. 7, but was moved, 1871, to Daley's store, SW sec. 7; thereafter it was kept in the village. It was disct. Aug. 15, 1902.

*** Peshugo Lake**

= *Hook L.* Possibly Winnebago; if so, a man's name. (Stucki) This is one of the distorted Indian names used only on Greeley's map of 1854; it gained little if any currency.

Peter's Prairie

For St. Peter's church, built there about 1859: the farmlands lying chiefly W and N of the crossroads settlement, *St. Peter's.*

Pheasant Branch

Another, probably the earlier, name for *Pheasant Branch Creek.*

Pheasant Branch, vill.

Probably for the creek; see next. The land was bought in 1838 by Thos. T. Whittlesey, of Conn., who laid out the village in 1849. It was platted 1853: secs. 1 and 12, MIDDLETON.

Pheasant Branch Creek

This is a later form of the original name Pheasant Branch,

evidently descriptive, there having been pheasants on this branch, or creek.

"The Pheasant Branch" was first used of the stream (earliest record found is 1845: Dane 127, 129, etc.). Next, the village of Pheasant Branch was laid out in 1849 and platted in 1853. Later, the pleonastic word "creek" was added to the name of the stream (earliest record found is 1861: Lig.), probably to distinguish it from the village, and also, no doubt, because "creek" is more familiar locally than "branch" (compare *Nine Springs Creek*).

Exactly what "branch" meant, in application to the stream, is a question. In the N Atlantic states it is not a common word, but when used, it means a division of a larger stream. In the S Atlantic states it is the most common word for a small stream, usually when this is a tributary, but even when it is the whole stream. (See *NED*, *DAE*, *Ling. Atlas of New England*, Wentworth's *Amer. Dialect Dict.*) Since Pheasant Branch (Creek) flows together with the water issuing from Whittlesey's Marsh before entering L. Mendota, it is possible that "branch" is here used in the N Atlantic sense; but it is certainly just as possible that it was given in the S Atlantic sense. The latter seems more likely because "branch" is coupled with "pheasant."

For, assuming that the name was descriptive, "pheasant" must have referred to some local bird. At the time when the name was given, the most common that could have gone by this name was the ruffed grouse (Schor.), called "partridge" by Northerners and "pheasant" by Southerners. Thus the evidence seems to point to the name's having been given by a Southerner.

Who he can have been is another question. Since Thos. T. Whittlesey bought a large part of this land, including the stream, in 1838, and laid out the village and named it in 1849, some have thought him also responsible for the name of the stream. But Whittlesey was from Connecticut.

However, Whittlesey was not the first in this location. In 1835 Col. Wm. B. Slaughter had purchased land just E of the mouth of the stream, where he planned to found his *City of the Four Lakes*. Since Slaughter was a Virginian, it seems more probable that he, finding what he would have called "pheasant" plentiful along what he would have called a "branch," named the stream (perhaps about 1835).

The stream rises in NE MIDDLETON and flows E into L. Mendota.

*** Pheasant Branch PO**

For the village. Est. June 28, 1850; disct. Dec. 31, 1902.

Pheasant Branch Spring

For nearby *Pheasant Branch*: a large spring in E sec. 1, MIDDLETON. (Schor.)

*** Pickarts' Corners**

For Gerhard Pickarts, storekeeper and postmaster here: the crossroads in W sec. 26, SPRINGFIELD. This name seems to have antedated the coming of *Ashton PO* to the corner (1867), and was probably used concurrently with "Ashton" for some years.

Pickerel Bay ['pɪkrəl]

For the fish to be caught there: an indentation in the shore of L. Monona (SW sec. 19, BLOOMING GROVE). So named on a plat of 1907.

Pickerel Point

See preceding entry: a point of land in SW sec. 19, BLOOMING GROVE. So named on Rural map, 1910.

Picnic Point

Because used as an outing spot from very early in the days of Madison's settlement: a narrow peninsula jutting northeastward into L. Mendota from the southern side; NW part of sec. 15, MADISON. The name appears first on Greeley's map, 1854.

Picnic Point Bay

For its being formed, in great part, by *Picnic Point:* a less common alternate name for *University Bay.*

Picture Rock(s)

Descriptive: the rocks have been eroded into striking shapes. This is a recent name (USGS, 1919–20, is first to record it) for the *Donald Rock*, not used in the locality; how it came to be on the maps is unknown.

Pierce's Hill ['pɪsɪz, 'pɪrsəz]

For J. G. Pierce, of N. Y., and family, who settled here shortly after 1842: a prominent hill in N secs. 3 and 4, OREGON.

Pierceville, vill. ['piərs,vɪl; 'pɝsə,vɪl (Kenn.)]

For Wm. A. Pierce (who became its first postmaster) and members of his family, who with others formed a settlement (1840) about the crossroads in NE sec. 26, SUN PRAIRIE. The name was probably used locally before the establishment of the PO.

*** Pierceville PO**

There has been some confusion about the form of this name. "Pierceville PO" was first est. Feb. 8, 1849, with William A. "Pierce" as postmaster (Official Register of the US). But in the Register for 1851, both the PO and the man are listed as "Pei-"; in 1853, both as "Pie-"; in 1855, both as Pei-"; from 1857 forward, both as "Pie-." Postal directories of 1851, 1855, 1856 and 1857 also give both as "Pei-." Park, however, spells "Pie-," and so the name appears on all the platbooks and maps. Thus the "Pei-" forms appear to be erroneous, and yet to have been in use for some years.

The PO ran till Jan. 16, 1861, when it was discontinued. It was re-est. as Pierceville PO, July 6, 1891, and ran so till its final discontinuation, Nov. 12, 1896. It was kept at *Pierceville*, vill.

Pierceville School

For the vill.; NE sec. 26, SUN PRAIRIE.

Pietchen's Pond [ˈpitʃɛ⊢nz]

For "Pietchen" (i.e., "little Pieter") Theisen, the former owner: a small pond in SE sec. 5, SPRINGFIELD. (Theisen was a small man.)

Pike Front, subd.

For the numerous pike caught here (Coll., Edws.); on L. Kegonsa, platted 1904; SW sec. 13, DUNN.

Pilgrim Village, subd.

A name intended to carry a favorable suggestion, with reference to the "pilgrims" of New England; to bear out the suggestion, the chief streets are named Standish Court and Alden Drive. Platted 1939, in SW sec. 21, MADISON.

Pine Bluff

Descriptive: a rocky ledge or bluff (running E and W thru N secs. 33 and 34, CROSS PLAINS) on which is a grove of mountain pines, a tree unusual in this part of the state. This was one of the earliest local landmarks, the road "From Blue Mounds to Four-Lakes" (Taylor 1838) running below and turning by it. The first recorded use of the name was by the PO, but it probably existed earlier.

Pine Bluff, vill. (also, **Pine Bluffs**)

For the nearby *Pine Bluff*. A small settlement about the crossroads between secs. 22 and 27, CROSS PLAINS; never platted.

Pine Bluff Church

For its location at Pine Bluff (NW sec. 27, CROSS PLAINS): St. Mary's Catholic church, built 1852–4, center of an early German settlement. (Barton 1103)

*** Pine Bluff PO**

For the nearby *Pine Bluff*. Est. Dec. 28, 1846, Geo. P. Thompson, who had land and a store in sec. 33, being the first postmaster. Disct. May 9, 1855. Re-est. Mar. 6, 1856; disct. Dec. 10, 1863; re-est. Apr. 25, 1864; disct. June 30, 1903. It was kept for a time in sec. 26 (Lig.) and later apparently in secs. 22 and 27. See also *Bluff PO*.

Pine Bluff School

For the bluff and the village. In NW sec. 27, CROSS PLAINS.

*** Pine Bluff Station,** vill.

For the Pine Bluff in adjoining CROSS PLAINS: a village, platted 1882 around the station of the C and NW railroad, in NE sec. 4, SPRINGDALE. *Bluff PO* was est. here the same year. The name was later changed to *Klevenville*.

*** Pine River**

For the pines along it. The name belongs to a river in Richland Co. which enters the Wisconsin R. from the W; thus its use for *Mounds Cr.* on Center's map of 1832 is an error.

Pines, The, subd.

For the name of the former farm, which had some striking rows of pine trees; platted 1909, in NW sec. 30, BLOOMING GROVE.

Plainview School

Descriptive (it has a view of a plain): at the road junction in SW sec. 32, PRIMROSE.

*** Pleasant Branch**

Erroneously used on some maps and in postal guides, etc., for *Pheasant Branch*.

Pleasant Grove School

Subjectively descriptive. In the extreme N of sec. 33, DUNKIRK, in the SW corner of the intersection.

Pleasant Hill, subd. (DUNKIRK)

Subjectively descriptive: the "hill" is no more than a slight rise in the ground; within the city of Stoughton, platted 1907; NE sec. 7.

Pleasant Hill, subd. (WINDSOR)

Subjectively descriptive; an addition to the village of DeForest, platted 1894; NE sec. 17.

Pleasant Hill School (DEERFIELD)

Subjectively descriptive: it is at the foot of an isolated hill in SW sec. 12. Formerly, *Balke School.*

Pleasant Hill School (DUNKIRK)

Subjectively descriptive. In S sec. 3. Formerly *Mennes School.*

Pleasant Site School

Subjectively descriptive. In NE sec. 21, MIDDLETON. Formerly *Lappley School.*

Pleasant Springs, town

Subjectively descriptive. "Pleasant Springs receives its name from a large spring on sec. 27, and numerous smaller ones in different parts of the town" (Butt. 911). Est. by separation from the town of Rome, Mar. 11, 1848: T6N,R11E. The Territorial Laws record the name, at the time of establishment, as "Pleasant Spring" but everywhere else the plural form is used.

Pleasant Springs School

For its proximity to the chief spring which gave the town its name. Near the center of sec. 34, PLEASANT SPRINGS. Now closed.

* **Pleasant Valley** (BERRY, etc.)

Subjectively descriptive. According to Park (599–600), this name superseded *Gorstville* and was superseded by *Farmersville.* But since "Farmersville" became official in 1848, and "Gorstville" was still current in 1849, "Pleasant Valley" can not have had much or long currency. If it refers to a specific local valley, this must be *Black Earth Valley.*

Pleasant Valley (OREGON)

Subjectively descriptive: another name for *Searls Valley.* (Ames)

Pleasant Valley (PERRY)

Subjectively descriptive: a valley stretching N and S thru secs. 10, 15 and 22. (USGS 1920–1) More lately, *Kittelson Valley.*

* **Pleasant Valley,** town

Subjectively descriptive. Park says, "The town was first called Gorstville; this was subsequently changed to Pleasant Valley, then to Farmersville, and then to Black Earth" (599). There is no doubt about the other three names, but Park's is the only evidence for this one.

Pleasant Valley Branch (of Pecatonica R.)

For the valley thru which it runs: a stream rising in sec. 3, PERRY, flowing S thru Pleasant Valley then W thru Drammen Valley, and into Iowa Co. (USGS 1920–1)

*** Pleasant Valley School**

For *Pleasant Valley* (PERRY): a former name for *Spring Valley School.* (USGS 1916–20)

Pleasant View, subd. (MADISON)

Subjectively descriptive; platted 1890, in NW sec. 26, MADISON; since absorbed into the city of Madison.

Pleasant View, subd. (WINDSOR)

Subjectively descriptive; an addition to the village of DeForest, platted 1898; SW sec. 17, WINDSOR. It includes part of *Campbell's Hill,* from which the view is presumably seen.

Pleasant View School

Subjectively descriptive: at the crossroads between secs. 8 and 17, BRISTOL.

Pleasure Park, subd.

Subjectively descriptive; on L. Monona; platted 1924, in NW sec. 25, MADISON. The name had been used for this locality long before the platting; Foote (1890) marks it so.

*** Point Mendota**

Descriptive: a point of land jutting into L. Mendota; former name of *Farwell's Point.* (Gree. 1854)

*** Point Monona**

From its location: a small point of land jutting into the NW side of Lake Monona, at the foot of Brearly St., Madison. Shown on early maps (Gree. 1854; Durrie 1856).

Pompey Hill (also, **Pompey's ——**) ['pampɪ]

Origin uncertain. Gov. W. R. Taylor, who owned the land and is agreed to have given the name, came here from N. Y. state, in which (Onondaga Co.) there are a town and village of Pompey, as well as Pompey Center; these may have suggested the name to him. (This, and other such classical names, were in vogue at the time. On his famous expedition with Lewis, Capt. Clark named a rock Pompey's Tower—now called Pompey's Pillar.)

There is a local tradition that Indians are buried on top of the hill, and this has given rise to another explanation (see the second form of the name) that an Indian chief named Pompey is buried there. But since there is no record of such a chief, this is probably a local manufacture.

The hill is in NE sec. 9, COTTAGE GROVE; Taylor settled here in 1848. It is also called *Old Pompey.*

Poverty Hollow

Ironic nickname of a small, fertile prairie in secs. 17, 18, 19, and 20, BERRY. (Butt. 924) The occasion of the naming is now unknown.

Prairie Addition, subd. (MAZOMANIE) ['prɛrɪ]

Because made on the prairie to the E of the village of Mazomanie; platted 1855, soon after the original plat of the village was made; SE sec. 9, MAZOMANIE.

Prairie Addition, subd. (WESTPORT)

Descriptive, but probably also having reference to the popularly accepted translation of *Waunakee* as "pleasant prairie"; an addition to the village of Waunakee, platted 1901; NW sec. 8, WESTPORT.

*** Prairie House, The**

For its location: a tavern owned by Horace Lawrence, of Vermont, in the 1840's and after, in about SW sec. 11, BURKE, on the *Token Creek Rd.* (Barton 1115)

*** Prairie Lake**

Descriptive: an early name for *Rice L.*, found only on the plat of *Troy* (1837), and therefore probably given by the platter, or his surveyor. Since Troy came to nothing, this name probably gained no currency.

Prairie Queen School

Named after the Prairie Queen Creamery, which was established at the same corners about 1904 (Onst., Reque): at the crossroads between secs. 9 and 10, CHRISTIANA. Formerly *Veeder School*, and *Rockney School.*

Prairie View School

Descriptive of its location on a slope overlooking flat open ground. Between secs. 35 and 36, FITCHBURG. Now closed.

Pratt's Hill ['prætsˌ]

For Ransom Pratt, of N. Y., an early owner of land (from before 1873 till after 1899) on which it chiefly is: a hill in S secs. 7 and 8, SUN PRAIRIE.

*** Prescott School** ['prɛs,kɑt, 'prɛskət]

For the Prescott family (see *Nora PO*), on whose land the school has been at sundry times. The platbooks locate it as follows: 1861, along the highway in S sec. 29; 1873 on Prescott's land, NW sec. 29; 1890, at Liberty Corners, in NW sec. 32 (across the highway from sec. 29), where it remained thru 1931; 1935 on Prescott's land again. It has also been called *Lee School*, and is now officially *Liberty School.*

*** Pridmore School** ['prɪd,mor]

For J. Pridmore, by whose land the school was established some time before 1873. Now *Pyburn School.*

Primrose, cor.

For the PO, which has 3 locations. Since the discontinuance of the PO, the village is indicated on different maps at its two most recent locations (Thrift, at the crossroads between secs. 16, 17, 20, and 21; Hiway, at the center of the town), tho at neither is there a settlement any more. At the latter, however, *Primrose Center School* still keeps the name.

Primrose, town

For the PO. The name "Perry" had been proposed, but since the PO was already "Primrose," the legislature gave this to the town too, and gave "Perry" to the town next W, est. at the same time (Mar. 21, 1849) by separation from MONTROSE. Thus T5N, R7E received a name originally chosen by Mrs. Spears. "Much talk was occasioned because the town was named by a woman." (Butt.)

Primrose Center School

For its location at the crossroad in the center of PRIMROSE. Formerly *Hanna School.*

Primrose Church

For its location: sec. 21, PRIMROSE; the common name for the "Primrose Norse Evangelical Lutheran Congregation" church, built 1855–6. (Park 527–8)

*** Primrose PO**

From a song which Mrs. Robert Spears, first woman settler of the town, had heard her father sing, beginning "On Primrose Hill there lived a lass." The PO, so named, had been awarded to her husband, and Mrs. Spears offered it also for the town, at a meeting to choose a name. Mrs. Chandler thought Primrose "too sweet," and proposed Hillsburgh instead, but the former was accepted (Barton). The PO, est. Apr. 27, 1847, was kept on sec. 19 by Spears till 1861, when it was moved to NE sec. 21; before 1890 it again moved, to SW sec. 16, where it remained till it was disct., Aug. 15, 1900.

Prospect Hill (MADISON)

Subjectively descriptive: a small hill on the SW shore of L. Monona, in NW sec. 25 (Foote, 1890, etc.). This was later platted as *Bellevue Park*.

Prospect Hill, subd. (WESTPORT)

Subjectively descriptive; platted 1940, in SE sec. 36.

Prospect Place, subd.

Subjectively descriptive, as having a view of L. Mendota; platted 1901, within the city of Madison; sec. 13, MADISON.

Pumpkin Hollow ['pʌŋkɪn ˌhɑlɒ˧, —— ˌhɑlə]

A semihumorous, but not derisive name, used from early days, for the fertile farmlands thru which the road runs for about 2½ miles N and S thru the center of BURKE.

Pumpkin Hollow School

For *Pumpkin Hollow*, in which it is, specifically, in extreme SW sec. 10, BURKE.

Putnam Hill ['pʌtnəm]

For John Putnam, who settled on it some time before 1861: a considerable hill over which the road goes in W-central sec. 22, OREGON.

Putz Landing [pʌts]

A spot along Waterloo Cr., in SW sec. 11, MEDINA, where an old settler named Putz used to keep his boat. It was a well-known fishing spot, and the name is still well known, tho not the location. (Wilt)

Pyburn School ['paɪbɚn]

For Thomas Pyburn, owner of the adjoining land from before 1890 till after 1911. The school is in NE sec. 24, SUN PRAIRIE. Formerly *Pridmore School*.

Quaker Heights, subd. ['kwekɚ]

For the adjoining Quaker Oats farm lands; probably also reminiscent; platted 1944, in SW sec. 9 and NW sec. 16, BLOOMING GROVE.

Quam's Park, subd. [kwɑ:mz]

For Ole J. Quam, owner of the land; on L. Kegonsa; platted 1896, in S sec. 25, DUNN.

*** Quam's Point**

For Ole J. Quam, who acquired the land some time between 1861 and 1873, and his descendants, in whose hands it remained until the 1930's: a former name for *Lund's Point*, and itself formerly called *Cedar Point.*

Quarrytown, subd.

For the three quarries about which a small early settlement was made. The quarrying had begun some time before 1861. It was platted 1863, including parts of NE sec. 20 and NW sec. 21, MADISON.

*** Quivey's Grove Precinct**

For its location at the house of Wm. Quivey, in S sec. 23, FITCHBURG: an election precinct, est. July 6, 1842, and including the present PRIMROSE, MONTROSE, OREGON, RUTLAND, SPRINGDALE, VERONA, FITCHBURG, and DUNN. It was superseded by the *Fitchburg Precinct.*

Railroad Bridge Island

Because it supports and is connected to the mainland by the railroad bridge: a long island in the Wisconsin R., below Sauk City; in the extreme NE corner of MAZOMANIE. (Derl.) The railroad came thru in about 1881.

Randall Court, subd. ['rændəl]

For adjoining *Camp Randall;* within the city of Madison, platted 1912; NE sec. 22, MADISON.

Randall Park, subd.

For adjoining *Camp Randall;* within the city of Madison, platted 1896; NE sec. 22, MADISON.

Rattlesnake Bluff

Because many rattlesnakes were found there formerly: a sharp, wooded bluff in W sec. 3, DANE. This was the early name; it is now usually reduced to "Snake Bluff." (Steele)

*** Ray,** town

Of uncertain origin. Park (594) mentions a James Ray as one of the English emigrants (1844), but since Kittle does not list him, he probably did not stay in the settlement. It is more likely that the town was named for Adam E. Ray, a member of the board of the Milwaukee & Mississippi RR, whose name (like those of other board members) was given to a street on the original plat (1855) of the village of *Mazomanie*. The town was organized out of BLACK EARTH, May 17, 1858, and included the S half of T8N,R6E; but it lasted so only till Jan. 15, 1859, when the area took back the name BLACK EARTH.

Raywood Heights, subd.

For "Ray" (Raymond R.) Frazier, owner, and his wife, whose maiden name was Wood (Woodw.); on L. Monona; platted 1900, in secs. 19 and 30, BLOOMING GROVE.

Red Brick School (ALBION)

Descriptive. In sec. 29, ALBION. Now closed.

* **Red Brick School** (MIDDLETON)

Descriptive: a former name for *Middleton Junction School.*

* **Red Bridge Crossing**

Descriptive? (See *Black Bridge Crossing.*) A crossing of the Yahara R. between L. Kegonsa and Mud L., in N sec. 14, DUNN. (Only on USGS, Madison Quadr. 1904, and Soils.) There had been an Indian ford here in early days (Dane 136).

Reider's Corners ['raɪdərz]

For F. Reider, who runs a store there: the junction of the town road from the N with highway 151, S sec. 28, BURKE.

Remy Spring ['rimɪ]

For August Remy and his descendants (who bought the land before 1890): the current name for *Grand* or *McFadden Spring,* in SW sec. 23, MONTROSE. (Fritz, Swig.)

Rice Lake [raɪs]

Said to be for the wild rice once plentiful there (now gone) but (see the neighboring *Sweet Lake*) it may have been for an early settler, Manson Rice (1843: Dane 71–2): a lake on secs. 13 and 14, ALBION. Shown on many early maps without name (Terr. 1835, etc.); first with the name on H&W, 1873, tho the name was probably in use for many years before that. Formerly *Prairie L.*, and recently, also *Bullhead L.*

Richardson('s) Cave

For David Richardson, who owned the land on which the entrance is, NE sec. 5, VERONA. The cave was discovered in 1842 by J. D. Sanford, and was formerly called *Cleveland's Cave;* it has also been known as the *Great Cave of Dane Co.* and *Verona Cave.* It runs beneath sec. 32, MIDDLETON. Richardson settled hereabouts in 1851.

Ridge, The

A ridge which separates the valleys of tributaries of Koshkonong Creek; it runs diagonally chiefly from secs. 14 to 12, COTTAGE GROVE, with the creek curving around it to the S, and the "Ridge Road" running on top. So called by all local residents since early settlement days. (W. and F. Kenn., etc.)

Ridge School

Because it is on *The Ridge*, and the Ridge Road; SW sec. 12 COTTAGE GROVE.

*** Riley PO**

For the village. Est. May 3, 1882; disct. about 1940.

Riley's, vill. (also, **Riley**)

For the platters of *Sugar River Station*, which is the official name of the plat, tho it was soon superseded by "Riley's" and "Riley Station," the current names today, since the C&NW railroad established its station there in 1882 under the first of these names.

Riley School

For the station and village about a mile N. The school is at the road junction between secs. 11 and 12, SPRINGDALE.

Riley Station, vill.

= *Riley's;* a frequent alternate name.

Rinden School ['rɪndən]

For Tollif Kittelson Rinden (or Renden), Norwegian settler (1843), owner of the land from before 1890, and his descendants, who remained there till after 1931. In NE sec. 9, PLEASANT SPRINGS.

Rinden School Hill

For Rinden School, which is on its N side: a sharp hill mostly in NE sec. 9, PLEASANT SPRINGS.

Ritchie School

For F. Ritchie and his family, who have owned the adjoining land from before 1873. In SW sec. 1, BURKE. Now closed.

*** River of the Four Lakes**

Descriptive: a former name of the *Yahara R.*, which flows thru the *Four Lakes.* Found first on Farmer's map (1828?), and continued at least as late as 1839 in Gov. Dodge's message to the Legislature. (Keyes 128)

*** River PO**

For its location by the Wisconsin R. in SW sec. 7, ROXBURY. Est. July 9, 1867; disct. Jan. 12, 1882.

Riverside Park, subd.

For its location along the Yahara River and Riverside Drive; probably also reminiscent; platted 1907 within the city of Madison.

Riverview School

Descriptive; from here the Yahara R. may be easily seen. SW sec. 21, DUNKIRK.

Robbins Creek

For the Robbins family (J., Z., and Z. M.) who settled along it before 1873, in secs. 21 and 28, COLUMBUS, Columbia Co. It rises in sec. 6, YORK, and flows immediately into Columbia Co., where it joins the Crawfish R.

Robertson's Grove

For David Robertson, early settler (1842) and owner of part of the land on which it is: a grove originally covering most of sec. 4, and adjoining parts of 3, 5 and 9, VIENNA (also part of adjoining Columbia Co.).

Rock Cut, (The)

Descriptive: a deep ravine in E sec. 31, MADISON, cut thru a rocky hill when the C&NW RR came thru here (1864).

Rockdale, vill.

Of uncertain origin; there were previous Rockdales in N. Y., Mass., Pa., etc.; however, it is in the valley of the Rock R., and the surrounding settlement was chiefly Norwegian, to whom at least the latter part of the name would appeal. It was platted 1885, including the former *Clinton,* whose name was superseded. It was incorporated 1914.

Rock Elm Park, subd.

Descriptive: rock elm trees grow there; on L. Kegonsa, platted 1902, in SE sec. 26, DUNN. The name was in use before the plat was made (Quam).

Rock House

Descriptive: a group of outcropping square rocks on a bluff in adjoining secs. 13 and 14, VERMONT. On the flat top of this bluff picnics and Fourth-of-July celebrations were held. (Pauls.)

* **Rockney School** [ˈrɑkˌnɪ]

For E. T. Rockney and his family, who held adjoining lands from before 1890 till after 1935. A former name for *Prairie Queen School;* it had itself superseded *Veeder School* (Onst.).

* **Rockside PO** (also, **Rock Side PO**)

Descriptive? Its exact location has not been discovered.

Est. June 10, 1857, with Andrew L. Mann as postmaster; disct. Mar. 31, 1860.

*** Rock Terrace,** hill

Descriptive: the rocky SE face of this hill, with outcropping strata, rises over 100 ft. in less than a quarter of a mile. Shown on Gree. map, 1854; NW sec. 26, BURKE. Renamed *Cincinnati Heights* about 1870.

Rocky Roost (Island)

Subjectively descriptive: a small, rocky island in L. Mendota, just N of *Governor's Island* (and joined to it since about 1940 by a strip of sand and vegetation). The island was formed when the lake level was raised (1850); the name appears first on a map in 1899 (Gay). The "roost" probably referred to a cottage built there by Robert Lamp and Mel Clark. (Sull.) Formerly, *Cat Island.* Also called *Lieutenant Governor's Island.*

*** Rome,** town

For Rome, Oneida Co., N. Y., at the suggestion of J. N. Ames. When first organized (Feb. 2, 1846) it included T5&6N,R9E, and T6N,R10E, i.e., the present OREGON, FITCHBURG and DUNN. When the first two of these were separately organized (Feb. 11, 1847), ROME was changed to include T6N,R10&11E, i.e., the present DUNN and PLEASANT SPRINGS. From this DUNN was separately organized (Mar. 11, 1848), and the name of the remaining township was changed to PLEASANT SPRINGS. Thus ROME became extinct.

*** Rome Corners**

As being the chief crossroads settlement in ROME: the former name of *Oregon* village. First settlement on the site (NW sec. 12) was in 1843; the village was platted and named Oregon in 1847, but the name "Rome Corners" continued in use at least as late as 1877.

*** Rome PO**

For Rome, Oneida Co., N. Y. Est. Mar. 9, 1846; changed to *Sun Prairie PO* Apr. 18, 1846.

Rosedale, subd.

Subjective, no doubt intended to suggest an attractive spot; however, it was not descriptive, since the land was a bare farm field. Platted 1917, in NE sec. 5, FITCHBURG.

Roundtop, hill (DANE) [ˈraʊnˌtap]

Descriptive: a large bluff mostly in NW sec. 10, DANE. (Steele) An early name, still fully current.

Roundtop, hill (MADISON)

Descriptive: it is somewhat domed; in adjoining parts of secs. 35 and 36, MADISON, rising about 100 feet above the surrounding lands.

Roundtop, hill (ROXBURY)

Descriptive: another name for the *Sugarloaf.* (Derl.)

Rowe's Spring [roz]

For R. W. Rowe, former owner of the farm on which it was: a spring now within the University Arboretum, at the W edge: SE sec. 28, MADISON. It flows N into the off-flow of *Big Spring,* and with it into L. Wingra.

Roxbury, town. ['raks,bɛrɪ]

For Roxbury, Delaware Co., N. Y., the birthplace of James Steele, an early settler, who suggested the name. The decision was made by vote, the alternative being "Nelson," suggested by Mrs. Burke Fairchild, in honor of the English admiral. The vote being tied, the secretary cast the deciding ballot. He spelled it "Rocksbury," however, and so it appeared on maps and elsewhere for a few years.

The town was est. Mar. 21, 1849, and included T9N,R7E. On Jan. 7, 1850, the part of T9N,R6E lying S and E of the Wisconsin R. was separated from *Farmersville* and added to Roxbury. When MAZOMANIE was organized, Jan. 15, 1859, it was given this fractional township, except secs. 13, 24, 25, 35, and 36, which remained part of Roxbury. On Feb. 27, 1865, however, these last sections were also given to Mazomanie, thus reducing Roxbury to its original area, which it has remained since.

Roxbury, vill.

For the town and PO: a village, never platted, along the road between secs. 16 and 21, ROXBURY. Just when the name was given is uncertain; settlement began when, in 1850, Father Inama gave some land in SW sec. 16 for a church and school. The PO was est. 1852, and perhaps kept here.

Roxbury Church

For the town: St. Norbert's German Catholic church, SW sec. 16, ROXBURY, nucleus of Roxbury village and German settlement; begun 1853.

*** Roxbury PO**

For the town. Est. Mar. 18, 1852, and apparently kept at or near the village in SW sec. 16; disct. Aug. 9, 1877; re-est. Sept. 4, 1878; disct. Aug. 15, 1902.

Ruddy Camp, subd. ['rʌdɪ]

For Peter Ruddy and family, who settled on this land about 1846; on L. Mendota; platted 1907 in NW sec. 27, WESTPORT.

*** Runey's Tavern**

For its proprietor, J. Bartley Runey, of Maryland, first settler of OREGON. The tavern, opened 1842 and named "The Pioneer's Hotel," stood in SW sec. 24 on the old *Lead Road.* Runey died in 1846, and the tavern was abandoned not long after.

Ruste School ['rʌstɪ]

For A. O. Ruste and his family, who owned nearby land (sec. 30) from before 1890 till after 1931. In NW sec. 29, BLUE MOUNDS.

Ruste Spring

For Johannes Ruste, an early Norwegian settler in W sec. 12, VERMONT: a spring beside the present highway 78; known in early days as a midway stopping place between Black Earth and Mount Horeb. It was also for a time called the *Helland Spring.* (Pauls.)

Rutland, town

For Rutland, Vermont, the early settlers of the W and SW part of the town having come from that state. Settlement began 1842; the town was est. Feb. 2, 1846: T5N,R10E.

*** Rutland,** vill.

Probably for the PO kept there, which was established in the same year as the town (1846): a settlement mostly at the E edge of sec. 19 at the crossroads, and partly in sec. 20; never platted; now called *Rutland Center.*

Rutland Branch

Because its sources are near *Rutland Center:* one of the branches of Badfish Cr.; formerly called *Anthony Cr.* This is a local name, not on maps.

Rutland Center, cor.

Since it is far from the geographical center of RUTLAND, this name probably refers to its once having been a center of settlement—as it was. This is the current name for the former *Rutland* village or *Rutland Corners.*

*** Rutland Center School**

For *Rutland Center;* it was in SE sec. 19, RUTLAND. Closed 1934.

*** Rutland Corners**

= The former village of Rutland, now *Rutland Center*.

*** Rutland PO**

For the town. Est. June 25, 1846; disct. Nov. 30, 1901. It was first kept at the village, but moved to extreme NE sec. 19 about 1880.

Saga Batomen ['saga 'batumen]

Norwegian for *Saw Mill Bottoms;* widely used in this strongly Norwegian settlement. (Barsn., Anders., Haugen)

*** Saint Joseph PO**

Perhaps for *St. Joseph's Church*? The exact location has not been discovered. Est. May 11, 1868, with Peter Kleiner as postmaster; disct. Apr. 12, 1869.

Saint Joseph's Church

A German Catholic church, center of *Die Sittlament*, NW sec. 11, BRISTOL. Built before 1861.

Saint Martin's Church

A Catholic Church, built 1886; center of the German settlement in S sec. 7, SPRINGFIELD.

Saint Peter's, cor.

For *Saint Peter's Church* here: a small crossroads settlement between secs. 27 and 28, SPRINGFIELD.

Saint Peter's Church

A Catholic church, in the center of a German settlement (*Peter's Prairie*); now in SE sec. 28, SPRINGFIELD. The original church was built at the same crossroads about 1859. Formerly also *Ashton Church*.

Sanderson School

For nearby *Sanderson Station*. The school is at the junction in NW sec. 19, BURKE.

*** Sanderson Station**

For Robert Sanderson, Secy. of the Madison and Portage RR, which built the line on which it was: a station, est. 1879, in SW sec. 17, BURKE; disct. 1897. A small settlement grew around it, but Butt. (932) is wrong in calling it a PO, since there never was such.

Sand Hill (DUNN)

Descriptive: a hill on which is the crossroads where secs. 19, 20, 29, and 30 meet. (Keenan)

Sand Hill (VERMONT)

Descriptive: a spur on the *Sand Ridge*, extending southward into the N part of sec. 26. (Deneen)

Sand Ridge

Descriptive. It runs thru secs. 8, 16, 15, 22, and 23, VERMONT.

Sandridge School (also, **Sand Ridge ——**)

For *Sand Ridge*, on which it is: in SW sec. 15, VERMONT.

*** Sandy Hollow School**

Descriptive of its location: a former name for *Searls School*, current in 1911. (Cant.)

Sauk Road, The [sɔk]

For Sauk City and Sauk Prairie, which it connected with Madison. Present State Highway 13 follows much of the original course, but the old road passed thru Pheasant Branch rather than Middleton, and went W from Springfield Corners thru NE BERRY (Barton 909, 924). It was in use from about 1844. See also *Upper Sauk Rd.*

Saunders Creek [ˈsɔndɚz]

For Jesse Saunders of Rensselaer Co., N. Y., who settled on it in 1842, on sec. 22, ALBION, now the site of the village of Albion, and who bought other land along the creek for several miles. The name first appears on the map of Foote, 1890, but was in use from the time when Saunders settled (see Butt. 838). It rises in sec. 20, CHRISTIANA, and flows S thru Albion into Rock Co. and Rock R. Formerly *Muskrat Cr.*

Saunders Run

For O. P. Saunders, and other members of the family, who have owned the land from before 1873: a short streamlet in N sec. 36, ALBION, running from a spring E into L. Koshkonong. (H&W, Clough)

Saw Bottoms

A recent variant of *Saw-Mill Bottoms;* or (more likely) an Englishing of *Saga Batomen.*

Saw-Mill Bottoms

For the sawmills along the stream which flows thru them: fertile alluvial lands in the valley of the E Branch of Mounds Cr., properly Elvers Cr., VERMONT. Sawmills were built here in 1847 (Sec. 21) and 1851 (sec. 28); they continued into the 1860's (Butt. 934). The bottom-lands are still referred to as the *Saw*

Bottoms, tho not all who use the name understand it; thus one informant (Linc.) called it "Sow Bottom," probably by folk-etymology.

*** Schaefer's Corners** ['ʃefɚz]

For J. D. Schaefer, who owned the land surrounding it from before 1861 till after 1873: a former name for *Lappley's Corners* (now *Nonn's Corners*).

Schaller School ['ʃælɚ]

For Albert Schaller and family, who owned adjoining land from before 1899: in NW sec. 5, MONTROSE. Formerly called *Malloy School*, and before that, *Griffith School*.

School Bluff

= *School Section Bluff;* a local variant.

School Section Bluff

From its being the chief feature of sec. 16 (MAZOMANIE) in which the old log schoolhouse known as *Howarth's School*, first in the settlement of the British Temperance Emigration Society, was built (1849). The land soon after changed hands, and the new owner would not give up the school building; so one night it disappeared and was found the next day on sec. 15, where the school has been since (Park 596). The bluff remained where it was, however.

Schoop's Lake [ʃops]

For C. Schoop and his family, owners from before 1873 till 1946: a small lake in SE sec. 18 and NE sec. 19, SPRINGFIELD.

*** Schroeder's Lake** ['ʃredɚz]

For Mathias Schroeder and his family, who owned the land off and on from before 1899 till after 1935: a former name for *Kalscheur's L.*, itself having replaced *Watzke's L.*

*** Schuetzen Park** ['ʃʊtzən]

From German *schuetzen*, "to shoot": it was used before 1877 by a local German club as a Shooting Park; in fact, Foote (1890) uses this English form rather than the German, which is used by Gay (1899), etc. It was along L. Monona in SE sec. 6 and NE sec. 7, BLOOMING GROVE, but was later platted as *Lake Park* and *Park View*, within the city of Madison.

*** Schumann's Lake** ['ʃumənz]

For the Schumann family (Lig., 1861, shows Christian and Fr[iedrich]) who owned adjoining land: an alternate local name for *Indian L.* (Dahmen). The family continued here till about 1900.

Scotch Lane

For the Scottish settlement there, and because at the E part, near the Presbyterian Church, Adam Davidson, a prominent settler, built high rail fences to keep other people's cattle off his land, thus giving the road the effect of a lane (Davids.): a local nickname for that part of present highway G running from W sec. 20, VERONA, to E sec. 25, SPRINGDALE. The name was in use from about the 1860's and is still well known, tho less used today.

Searls School ['sɝlz]

For the Searls family (see *Searls Valley*). In NW sec. 33, OREGON. Now closed.

*** Searls Valley**

For Peter, John, S., and M. W. Searls (or Searles), who, some time before 1861, settled on adjacent farms in N secs. 32 and 33: a valley in NW sec. 33, OREGON.

Second Lake

An early name for *L. Waubesa.*

Second Point

For its location W of Madison along the S shore of L. Mendota (Picnic Point counting as the first): in S sec. 9, MADISON. So called from before 1900. (Mendo., Brown)

Second Ward Beach, subd.

Because the lots were bought first by people living in the second ward, Madison; on L. Mendota; platted 1909, in SE sec. 28, WESTPORT. The name, tho still on the plat, has fallen into disuse. See *North Shore Bay.*

Section Bluff

An abbreviation of *School Section Bluff.* (Kittle, 82)

Seminary Springs, cor.

For the Seminary Springs Farm adjoining it. The farm name was registered by the owner, F. G. Good, on Aug. 10, 1915, as a means (it is said) of preventing its use by others in the vicinity. It had been used by Good for many years before: the platbook of 1911 (Cant.) shows "Seminary Spring" on his farm, in the extreme NW corner of sec. 6, and the farm had been in possession of the Good family since before 1873. The springs are in BURKE, SUN PRAIRIE, and COTTAGE GROVE; they are the source of DOOR Cr.

The name refers to the fact that the lands were originally among those set aside for the support of the University under Act of Congress (June 12, 1838) "An act concerning a seminary of learning in the Territory of Wiskonsan."

The corner, a junction of three roads, is in extreme SE sec. 36, BURKE, with stores and the school, but the settlement extends SE into COTTAGE GROVE.

Seminary Springs School

For *Seminary Springs* corner, at which it is. Formerly *Walbridge School.*

*** Seven Mile Creek**

= *Six Mile Creek;* apparently an error (Butt. 870).

Seventy-Six Farm, The (always written "The '76 Farm")

From the sign erected about 1840 on his farm (on secs. 27 and 28, BURKE) by Alexander Botkin, in reference to the year of the Declaration of Independence. Tho the land has changed hands many times, the name is still in local use.

Shady Side Park, subd. (also, **Shadyside** ——)

Descriptive: it has many trees; platted 1892, on L. Kegonsa; SW sec. 30, PLEASANT SPRINGS. The name was used from before 1890 (Foote).

Shady Lawn School

Descriptive: it is on a grassy spot among trees. E sec. 25, YORK.

Shepard's Marsh (also, formerly, **Shephard's** ——)

For Luther G. Shepard, of Mass., and his family, owners since 1845: a large marsh in SW YORK, especially secs. 19 and 30; recently reduced in size by drainage.

Sherman Park, subd.

For Sherman Ave. (itself named for the General), at which it begins; within the city of Madison, platted 1927.

Sholts's Hill [ˈʃoltsɪz]

For J. Sholts and family, owners of the adjacent land (SE sec. 4) from before 1911: a prominent hill in S sec. 3, OREGON. Formerly *Underwood Hill.*

*** Shooting Park**

= *Schuetzen Park.* The Anglicized form is found in Foote's platbook, 1890, but the German form was usual.

Shore Acres, subd.

For its location on the shore of L. Monona; platted 1911, in NE sec. 17, BLOOMING GROVE.

Shore Acres Lagoon

A small inlet on the shore of L. Monona in the S part of *Shore Acres.*

Shorewood, subd.

Descriptive; along L. Mendota; platted 1922, in E sec. 17, MADISON. It absorbed parts of *Merrill Park* and *Mendota Heights*, and was itself absorbed later into the village of *Shorewood Hills*, tho the name "Shorewood" remains in general use for this village.

Shorewood Hills, vill.

A combination of the names of the two main subdivisions which composed it: *Shorewood* and *College Hills.* It became an incorporated village in 1927, and now includes parts of secs. 16, 17 and 21, MADISON.

Showers School ['ʃaʊɚz]

For Samuel Showers and his family, who have owned land in secs. 23 and 24, CROSS PLAINS, from before 1847. The school is in E sec. 24; also called *Walters School.*

*** Silver Lake**

Generously descriptive, since the only water at the site was known later, and more accurately, as *Mud Lake*: the lake depicted on the plat of the paper town *Middletown* (1837). Its size and shape were altogether imaginary; actually, Mud Lake covered much of the southern part of the plat, which was supposed to surround Silver Lake. This name can hardly have gained much currency.

Silver Spring

For the clarity of its water: a source of drinking water for Madisonians for many years; the name is shown in Cantwell platbook, 1911, but existed long before. NE sec. 34, MADISON.

*** Sittlament, Die** [di ˌsɪtlɑ⊥'mɛnt]

A Germanized form of "the Settlement": an early name among German settlers for the region of East Bristol, where their nationality was concentrated. (Renk)

Six Mile Creek (also, **Sixmile** ——)

Descriptive of its approximate length: a creek rising in NE SPRINGFIELD, with branches in adjoining corners of VIENNA and DANE; it flows generally SE thru WESTPORT into L. Mendota, sec. 28. First appearance of the name on a map was in 1861 (Lig.). Formerly *Neosho Cr.*

Skellyville, vill.

For Leo Skelly, owner of the land: a small collection of dwellings mostly in NE sec. 23, FITCHBURG (Purc., Barry). The name is local and unofficial. The allotment was made after 1931.

*** Skeneda Creek**

= *Door Cr.* Perhaps Winnebago; [skɛnida] = "the pure water" (Stucki). This is one of the Indian or pseudo-Indian names used only on Greeley's map of 1854; it gained little if any currency.

Skunk Hollow

Because of the many skunks there: a former name for *Pleasant Valley*, OREGON; not considered derisive in the locality. (Ames., Cong.)

*** Slaughter's Marsh**

For George and William B. Slaughter, who owned the N part of it (S sec. 3) from before 1873 till before 1890 (see *Peatville*): an earlier name for the *Big Marsh*, MIDDLETON.

Smithback School

For Nils Olsen Smithback, from Norway, who settled in 1842 on the adjoining land, and remained till after 1873. SW sec. 15, CHRISTIANA.

*** Smithfield,** town

Origin unknown; if for some earlier place of the same name, there were aready at least six in the U. S. Shown only on Chapman's map, 1855, instead of *VERMONT*. Since VERMONT was established and named in that year, it may be that the mapmaker, hearing that the new town was to be called Smithfield, recorded that before the actual decision was made. See *HOBART*.

Smith School

For J. W. Smith, who owned adjoining land from before 1873 till after 1911. S sec. 26, MEDINA.

*** Smoky Mountains,** hills

From the bluish haze or fog often seen near their tops, and which was also responsible in part for their present names, the *Blue Mounds*. The name is said to have originated with the Indians, tho their form of it has apparently not been recorded. H. Huebbe's map (Gotha 1825) shows "Smocky Mts." The term was still in use in 1876. J. R. Brigham says this name appeared on the "old school atlases" (Park 239–40), but these have not been traced.

Snake Bluff
= *Rattlesnake Bluff.*

*** South Branch (of Sugar River)**
Because it is S of the main branch: a former name for the present *West Branch;* so called on maps of 1855 (Chap., Gree.), and the platbooks from Foote (1890) to the present.

South Madison, subd.
For its position with respect to the city of Madison, of which it was a suburb when platted, 1889; N part of sec. 26, MADISON. Since absorbed into the city of Madison.

South Park, subd.
For its location on S. Park St.; a subdivision of *Greenbush;* platted 1908, and with it later absorbed into the city of Madison.

Sow Bottom [ˈsɑʊ ˌbɑtəm]
For *Saw Bottoms* (Linc.); probably by folk-etymology, thru some association with pigs.

Spaanem Creek [ˈspɑnəm]
See *Spaanem Hill.* This is the E branch of what Highway maps call *Deer Creek.* The name is little used today. (Gilb.)

Spaanem Hill
For Thore Spaanem, who settled (1846) by the spring at its foot in NE sec. 17, from which Spaanem Creek rises: the south part of what is called on its N half *Berg Hill,* S sec. 8 and N sec. 17, SPRINGDALE. (Spaan.)

*** Spears Settlement, The** (also, erroneously, **Speers**) [spirz]
For Robert Spears and family; see *Spears Valley:* a name in use in 1848 (on a Ms. map of a road; Dane Co. Court House).

*** Spears Valley**
For Robert Spears and family, from Ohio, first settlers of PRIMROSE (1844): the valley of the creek flowing from their "big spring" on sec. 19, N thru sec. 18, and NE thru 17, thru which, also, went one of the earliest important roads in the town. (Prim. 34)

Speedway (Road), The
Because when newly surfaced it was used by early autoists as a speedway: the common, tho unofficial, name for the part of *Mineral Point Rd.* between Madison and Pine Bluff.

Spring Brook
The current name for *Spring Creek,* MEDINA. (Soren., Wilt)

* **Spring Creek** (BERRY, etc.)

Because fed by springs: a tributary of Halfway Prairie Cr., which it joins on sec. 16, MAZOMANIE. It rises in sec. 15, BERRY (USGS) and flows W. Shown on most maps, but named only by H&W, 1873. This may be no more than a description; the creek has no local name today.

Spring Creek (DANE)

For its origin from many full-flowing springs. It rises in the NW part of DANE and flows NE into Columbia Co. Shown on maps from 1849 (Lapham).

* **Spring Creek** (FITCHBURG)

Mentioned in Butt. (869), but of uncertain identity, tho said to be in FITCHBURG. It may be *Murphy's Cr.* or perhaps *Nine Springs Cr.*

Spring Creek (MEDINA)

Because it originates from springs: a creek rising on sec. 22, MEDINA, and flowing N to enter Waterloo Cr. on sec. 15. (H&W, 1873, etc.) The USGS map (1905) shows it as draining from sec. 34, but the springs are the actual source.

Spring Creek (SPRINGFIELD, WESTPORT)

Because fed by springs: a creek rising near the center of SPRINGFIELD, and flowing E into WESTPORT, where it enters Six Mile Cr. on sec. 28. First named in Cant. (1911). Also called *Dorn's Cr.*

Springdale, town (formerly also, **Spring Dale**) [ˌsprɪŋ'del]

In part, at least, descriptive, since it has many springs and valleys; it may also be partly reminiscent, since there was at least one other Spring Dale in existence, that in Hamilton Co., Ohio (tho there is no evidence that there were Ohioans among the earliest settlers). There were, however, Norwegians among these settlers, and they may well have had some word in the choice, "spring" and "dale" having cognates very frequently found in Norwegian place-names. The town was organized in 1848, to include T6N,R7E.

Springdale Center School

For its location a quarter mile N of the center of SPRINGDALE.

* **Spring Dale PO** (also, unofficially, following the name of the town, **Springdale PO**)

For the town. Est. Apr. 9, 1850, and kept by Thomas B. Miles on sec. 25; disct. Dec. 11, 1890.

Springfield, town.

Descriptive and reminiscent? If named for a settlement farther E or S in the U. S., there were already 20 or more by this name. It became a separate town (T8N,R8E) Apr. 6, 1847.

Springfield Corners, vill. (also, **Springfield**)

As being the chief settlement in SPRINGFIELD (SE sec. 5); but whether the settlement took this name from *Springfield Corners PO*, or the PO took the name from the settlement, is uncertain. The latter seems more likely. Formerly *Clark's Corners;* also *Halunkenburg.*

*** Springfield Corners PO**

For the town, and perhaps the settlement at which the PO was kept. The name was changed from "Dane PO," Nov. 24, 1871; it was disct. July 31, 1903.

Springfield Corners School

For the nearby *Springfield Corners.* NW sec. 9, SPRINGFIELD.

Springfield Hill

For Springfield Corners, the settlement nearest to it: the longest and highest hill over which the Madison-Sauk road passes, therefore important to travellers from early days. It is mostly in SPRINGFIELD and ROXBURY, but also BERRY and DANE, where the four towns meet. Also, *Lutheran Hill.*

Spring Harbor, subd.

Because it includes *Merrill Springs* and the mouth of *Warner Creek,* widened to form a small harbor; on L. Mendota; platted 1910, in SE sec. 18, MADISON.

Springhaven, subd. (also, **Spring Haven**)

For the Springhaven farm, on a corner of which it was platted, 1916; in NE sec. 17, BLOOMING GROVE. The "springs" are actual; the "haven" was no doubt subjective.

*** Spring Lake**

Because fed by springs: a pond in SW sec. 17, FITCHBURG. Shown and named only by Foote, 1890.

Spring Park, subd.

For the nearby *Merrill Springs;* platted 1892, in SE sec. 18, MADISON.

*** Spring PO**

For *Spring Valley*, or the spring which gave it its name? Est. May 5, 1896; disct. Nov. 30, 1900. The postmaster was Gottlieb Grunder, and it was kept in or near the N part of sec. 10—the head of Spring Valley—PERRY.

Spring Slough, The ['sprɪŋ ˌslu]

Because, having a spring, it does not freeze over (cp. *Ice Slough*): a backwater from Blum's Creek, just E of Ice Slough, in SW sec. 7, ROXBURY. (Derl.)

Spring Valley

From the Spring Valley Cheese Factory, founded here some time before 1890. The name is descriptive: the upper valley of *Pleasant Valley Branch*, chiefly sec. 10, PERRY, where the stream rises from springs.

*** Spring Valley School** (BERRY)

For its location in the valley of *Spring Creek*. A former name for *Danz School*.

Spring Valley School (PERRY)

For *Spring Valley*, at whose head it is; S sec. 3, PERRY. Formerly *Pleasant Valley School*.

Squaw Bay

For *Squaw Point*, enclosing it to the W: an inlet in the E shore of L. Monona, in W sec. 20, BLOOMING GROVE.

Squaw Hill

For some association with the Indians, no longer known: a considerable hill in NW sec. 16 and NE sec. 17, WESTPORT.

Squaw Hill School

For *Squaw Hill*, to the N of it. W sec. 16, WESTPORT. Now closed.

*** Squaw Point**

Probably for the wife of Abraham Wood (see *Wood's Point*), a daughter of the Winnebago Chief Decorra. The word "squaw" is Algonquian, not used by the Winnebago, but a common adoption of the whites in the East; thus it was very likely brought here by some white person, very likely Wood. Since there had been a Winnebago village here before the whites began to settle, it is possible that the name may have a general reference; however, the presence of a particular "squaw," a chief's daughter and wife of a white man, seems to give better grounds for the name. See also *Wood's Point* and *Winnequah*.

Staangji ['stɔŋɪ, 'stɔŋɛ]

Norwegian, meaning "the pole": a name (still widely known) among the local Norwegian settlers for the crossroads about which the village of *Mt. Horeb* began. It referred to the presence there of a flagpole, which the English-speaking settlers called the Liberty Pole. (Gilb.)

Stapelman's Lake ['stɑplmənz]

For the Stapelman family (H&W, 1873, shows Jno. Stapelman), whose members have long owned adjacent land: an alternate local name for *Indian L.* (Dahmen)

Starkweather('s) Creek ['stɑrk,wɛðɚz]

For John C. Starkweather (later General) who, as Sergeant at Arms for the Convention of 1846, was bringing a payroll from Milwaukee to Madison, and, unable to cross the creek E of the village, felled some trees and made a bridge, which remained in use for some time and gave the creek his name (Stark.). One map (Monona 1900) places the name on Burke Cr., but it is properly the stream rising in sec. 27, BURKE, and flowing SW thru BLOOMING GROVE into L. Monona. Another early name was *Clyde Cr.*

Starr Church [stɑr]

For the Starr family, early members of its congregation (organized about 1845), and the Rev. D. F. Starr, its first pastor: the First Free-Will Baptist Church of Rutland, built 1874 at what became *Starr Church Corner.*

Starr Church Corner

For the *Starr Church:* the crossroads where secs. 1, 2, 11, and 12 meet, RUTLAND.

Starr School (also, **Star ——**)

For the Starr family (see *Starr Church*); the original school was built before 1861 at the present *Starr Church Corner.*

Starvation Hill

Origin unknown, but it sounds like a jocular name: a hill mostly in NE sec. 31, DUNKIRK. (Johns.)

Steamboat Rock

Because it resembles a steamboat: a large outcrop of rock at the top of the bluff in NE sec. 18, DANE. An early name, still in some use.

Steele School (DANE)

For James Steele, of N. Y., and his descendants, owners of the adjoining land since 1848. The school is in SW sec. 9. Now closed.

*** Steele School** (VERMONT)

For Isaac C. Steele, of N. Y., who settled on adjoining land in 1848, and remained till after 1880. A former name for *Booth School.*

*** Steele's Creek**

For Robert Steele, a trapper, who had settled on it at least by 1832: an early name for *Black Earth Cr.*, at least where Steele was settled on it, near Cross Plains (Dahmen).

Steensland, subd. ['stinzˌla⊥nd]

For Halle Steensland, owner; platted 1891, the first plat made at Maple Bluff; NW sec. 1, MADISON. Since absorbed into the village of Maple Bluff.

Steensrud School ['stɛnzˌrʊd, 'stinzrʊd]

For Ole A. Steensrud, of Norway, who settled on adjoining land before 1873, and for members of the family still in the town. In extreme SE sec. 4, VERMONT.

Step Valley

For the Step Valley Cheese Factory, i.e., the valley had no name until the factory was to be named (some time before 1890). "Step" is said to be somebody's name, but since no trace of any such person has been found, the name may be descriptive. It is used on the USGS map, 1920–1, which shows part of the valley in sec. 32, PERRY, tho it is mostly in Green Co. The factory burned, and the name is now virtually forgotten in the locality.

Stewart Park ['stuɚt]

For Frank Stewart, Verona, who, as Chairman of the County Board, was influential in its construction in the late 1930's: a park containing an artificial lake, secs. 1, 2, and 12, BLUE MOUNDS.

Stone, cor.

For the PO, and ultimately the school: a crossroad settlement where secs. 21, 22, 27, and 28 meet, RUTLAND.

*** Stone College**

Subjectively descriptive: a former local name for the *Meffert School.* It was built of stone, and the word "college" referred to "its bright pupils and the high grade of work done there" (Barton 1035).

*** Stone PO**

For the *Stone School.* Est. June 10, 1895, and kept across from the school at the crossroads, now *Stone* corners; disct. Mar. 30, 1901.

*** Stone Quarry Hill** ['ston ˌkwɔrɪ]

Descriptive: there was a stone quarry here in early days (shown by H&W and Foote): a small hill in SW sec. 30, DUNKIRK. (Dow)

*** Stone Quarry Point**

Descriptive: an early name for *Macbride's Point.* (Park 133)

Stoner School

For *Stoners Prairie PO,* which was nearby, and ultimately the Stoner family. At the road junction between N secs. 19 and 20, FITCHBURG. Now closed.

Stoner's Prairie (also, **Stoners** ——, and simply, **Stoner's**)

For John Stoner and his family. Stoner opened (tho he did not settle) the first farm in FITCHBURG (1838) on this prairie, which covered parts of secs. 17, 18, and 19, and the adjoining secs. 13 and 14, VERONA.

*** Stoners Prairie PO**

For the prairie. Est. Jan. 24, 1851, disct. Feb. 25, 1885; re-est. Mar. 6, 1886, disct. Apr. 13, 1887. Kept in the NW part of sec. 20, then in the N part of sec. 30, FITCHBURG.

Stoner's Prairie Road

Because it passes thru *Stoner's Prairie;* a less common alternate name for the *Fitchburg Rd.* (Park 448)

Stone School

Descriptive: when built (before 1861) it was the only one of stone in the town. At the crossroads where secs. 21, 22, 27, and 28 meet, RUTLAND.

Storre Spring ['sto⊥rɛ⊢]

Descriptive, meaning "large" in Norwegian: a spring in NW sec. 34, DEERFIELD, flowing into Mud Cr. from the E. The same name was also used of *Borghilda Spring* (Reque). No longer much used, as less Norwegian is spoken in this region.

Story, cor.

For the Story family; see *Story Creek:* a small settlement along the road between secs. 18 and 19, OREGON. The name was probably in use, in this form, and certainly in the form *Storytown* before the establishment of the PO (1890).

Story Creek

Ultimately for L. M. Story, T. Story, and their families, who settled (1846–7) in secs. 18 and 19, OREGON, at the head of the creek. The stream appears on maps, until recently, without name, but this name was probably used locally much earlier; cf. *Story, Story PO*, and *Storytown*.

*** Story PO**

For the *Story* settlement at the head of Story Creek. Est. Jan. 8, 1890; disct. Oct. 31, 1903.

Story School

For the *Story* settlement. In SW sec. 17, OREGON.

Storytown, cor.

= *Story*, cor. This was the common name of the settlement in early days, and is still used by "old-timers"; the PO name of "Story," however, is the current form today.

Stoughton, city ['stotən, 'stotn̩]

For Luke Stoughton, of Weathersfield, Vt., its founder. Platted, and first settlement made 1847; became a village 1868, a city 1885; it now includes parts of secs. 4, 5, 6, 7, 8, and 9, DUNKIRK.

Stoughton PO

For the village (now city) of Stoughton. Est. Oct. 25, 1848; the name and location were changed to *Bass Lake PO*, Apr. 9, 1850, but a new Stoughton PO was est. Mar. 14, 1851.

*** Strawberry Point** (BLOOMING GROVE)

From the wild strawberries once plentiful there. Another name for *Wood's Point.*

*** Strawberry Point** (MADISON)

Because wild strawberries were plentiful there: a former name for *Picnic Point.* (Brown)

*** Straw Point**

Apparently an error (in Butt. 927) for *Squaw Point;* but since another name for this place was Strawberry Point, the error was not without some point.

Sugar Bush, The (also, **Sugar Bush Grove**)

Descriptive: a large grove of sugar maples on the farm of J. I. Williams, covering parts of secs. 18 and 19, PLEASANT SPRINGS; a well-known picnicking spot from at least the 1870's.

Sugar Bush Point

For *The Sugar Bush*, which is partly on it: a point of land jutting half a mile into L. Kegonsa from the NE shore; W sec. 19, PLEASANT SPRINGS.

Sugarloaf, The

For its more or less conical shape: a striking, solitary hill, standing out from the E bluffs of the Wisconsin R., NW sec. 6, ROXBURY. (Derl.) The name (commonly applied to this type of hill) probably goes back to early settlement days, when sugar came in conical loaves. Also called *Roundtop*.

Sugar River (also, early, **Sugar Creek**)

From the abundance of sugar maples near its mouth. "Tonasookarah," generally given as the Winnebago name, is said to mean "sweet or sugar water," but this cannot be the exact form. It may represent either *da ni shu o gu ra*, "flowing from or thru a sugaring place or sugar maples" or *da ni shu o ka ra*, "the sugar giving" (Stucki). "Sugar River" is evidently a rough rendering of one of these into English. (Another explanation, that the name was given in 1833 by the government suveyors, who were delighted with the sweetness of the water, is clearly disproved by the appearance of the name on maps as early as 1829 (Chan.).)

The stream has many branches, which has led to much confusion and contradiction in the names. The present *West Branch*, for example, was apparently taken as the main one (Center, 1832), and *Badger Mill Cr.* was taken (1844) as the "head waters" (Dane 90). However, there has been general agreement from the first that the main branch, or "Sugar R.," is (as at present) that which rises in sec. 32, CROSS PLAINS and flows E, then turns SE across the corner of SPRINGDALE and most of VERONA into MONTROSE, on sec. 28 of which it is joined by the *West Branch;* it then continues SE into Green Co., where it is joined on sec. 21, ALBANY, by *Little Sugar R.;* thence it continues thru Green Co., and SW Rock Co., into Illinois, where it joins the Pecatonica R.

Its former names have been *Muddy R.* and *Pectanon R.*, applied early to the lower part of the stream. The upper part was also called *Eastern Branch*, *North Branch*, and *Sutherland R.* See also other former names of branches: *Middle Branch* and *South Branch.*

*** Sugar River Station,** vill.

For the nearby *Sugar R.;* platted 1881 by Robert and William Riley, in SE sec. 2, SPRINGDALE. See *Riley's*.

Sugar River Valley

For the *Sugar R.*—a broad valley including parts of CROSS PLAINS, SPRINGDALE, VERONA, PRIMROSE, and MONTROSE and parts of Green Co., etc.; having fine farm land, it was the site of many early settlements.

Sumac Lagoon ['su,mæk lə'gun]

For the sumac plants growing there: the inmost channel of *Belle Isle.* Named on the plat, 1928, by L. S. Davis. (Owen)

Summit Park, subd.

For the nearby *Summit Station;* platted 1911, in NW sec. 32, MADISON.

Summit Ridge, subd.

For its position (cp. *Summit Park);* platted 1925, in SW sec. 32, MADISON.

Summit Station

For its location on high land overlooking Madison: a station on the IC railroad, est. 1888, in NW sec. 32, MADISON.

Sunny Knoll, subd.

Subjectively descriptive; platted 1935, in N sec. 5, BLOOMING GROVE.

Sunnyside, subd.

Subjectively descriptive (it is on the E side of greater Madison); platted 1911, in adjoining parts of secs. 29, 32, and 33, BURKE.

Sunnyside Heights, subd.

Because it lies higher than *Sunnyside,* to which it is an addition; NE sec. 33, BURKE. Platbooks of 1926 and 1931 show it, tho it was never platted under this name.

Sunnyside School (BURKE)

For nearby *Sunnyside.* In SW sec. 28.

Sunnyside School (CHRISTIANA)

Subjectively descriptive; named about 1905. At the crossroads between secs. 30 and 31. Formerly *Vee School,* and after that, *Hulsether School.*

Sunnyslope, subd.

Subjectively descriptive; platted 1913, in NW sec. 35, MADISON.

Sunnyslope School

Subjectively descriptive. In NW sec. 12, YORK.

Sun Prairie

So named on June 5, 1837, by Augustus A. Bird, Acting Commissioner for the erection of the capitol at Madison, and 45 men accompanying him, who, having come westward from Milwaukee through cloudy and rainy weather for 9 days, came upon this prairie in bright sunshine. The name was at once carved on an oak tree nearby to mark the event. Originally the prairie covered the greater part of the present SUN PRAIRIE and the W side of adjoining MEDINA.

Sun Prairie, town

For the prairie and village. When first organized, 1846, it included T8&9N, R11&12E; on Mar. 4, 1848, however, the present BRISTOL, MEDINA, and YORK were separately established, reducing SUN PRAIRIE to T8N,R11E.

Sun Prairie, vill.

For the prairie. The first settlement made in the town (by C. H. Bird, 1839), about two miles W of the point at which the prairie had been discovered and named, became the nucleus of the present village. It was platted and incorporated 1868; it now includes parts of secs. 4, 5, 8, and 9, SUN PRAIRIE.

Sun Prairie-Columbus Road, The

An alternate name for the early *Columbus Rd.*

*** Sun Prairie-Lake Mills Road**

For two main settlements on it: an early road, also called the *Aztalan Highway.*

Sun Prairie PO

For the village and town. The name was changed from *Rome PO*, Apr. 18, 1846.

*** Sun Prairie Precinct**

For *Sun Prairie*, nearby: an election precinct, est. Apr. 5, 1842. In 1844 it included the present YORK, MEDINA, SUN PRAIRIE, BRISTOL, and northern two-thirds of WINDSOR. (Dane)

Sun Prairie Road, The

Because it joins Madison to Sun Prairie: so called as early as 1846 (Dane 148). The present course (US highway 151) is only slightly changed.

Sunset Hills, subd.

Descriptive, and probably influenced by the neighboring Sunset Village, Sunset Ridge, etc.; platted 1941, in SW sec. 21, MADISON.

Sunset Point

Because it is and has for many years been known as an excellent point from which to observe the sunset: a hilltop on the W edge of sec. 21, MADISON. The name has been in use for 60–70 years. (Woodw.)

Sunset Ridge, subd.

Descriptive and probably influenced by Sunset Village, to which it is an addition; platted 1941, in NE sec. 20, MADISON.

Sunset View Park, subd.

Presumably descriptive (it is E of Madison whence one can see the sunset across L. Monona); platted 1915, in NE sec. 8, BLOOMING GROVE.

Sunset Village, subd.

Subjectively descriptive, referring to *Sunset Point*, below which it lies, and also referring to the fact that J. C. McKenna, Sr., who named it, was planning to retire from the real estate business, and expected this to be his last plat, the "sunset" of his work; platted 1938, in SE sec. 20, MADISON. (McKen.)

*** Superior**

Probably subjectively descriptive and reminiscent; some reference may be had, however, to the bluffs of the Wisconsin R., along which it was: a "paper town," platted Feb. 28, 1837, in secs. 6, 7, 18, and 19 of the present ROXBURY, and fractional sec. 13 of the present MAZOMANIE. Tho shown on early maps (Mitchell 1838, etc.) it never came to anything.

*** Sutherland Hill**

For Chester Sutherland, of N. Y., and his family, who owned land here from 1837 till after 1911: another name for the *Nine-Spring Hill*, of which this is the S side. (Brown 7)

*** Sutherland River**

Origin uncertain. Used only on Cram's map (1839) and Doty's (1844), for the upper part of *Sugar R.* Perhaps named for Thomas W. Sutherland, who in 1835 floated in a skiff down the Mississippi from the Falls of St. Anthony to the mouth of Rock R., then paddled up that stream and the Yahara R. to the present site of Madison, where he settled.

Swan Creek

For the presence of swans on it? On the other hand, it flows into *L. Waubesa*, which means "swan lake." The creek rises in secs. 11 and 22, FITCHBURG; the branches meet in sec. 13 and flow SE thru DUNN to the lake (USGS). Tho the name is not recorded till 1880 (Gutt.), it must have been in existence before that, particularly if there is any connection with the naming of L. Waubesa.

Swan Creek School
For nearby *Swan Creek*. In N sec. 13, FITCHBURG. Now closed.

*** Swan Lake**
An early name for *L. Waubesa.*

*** Sweet Home PO**
Reminiscent of the song "Home, Sweet Home"? Est. May 9, 1855, on the land of its first postmaster, Wm. H. Miller; NW sec. 31, DEERFIELD. Disct. Nov. 14, 1861.

Sweet Lake (also, **Sweet's ——**)
For Freeborn Sweet of Otsego Co., N. Y., who settled beside it in 1841: a small lake, now virtually dried up, mostly on sec. 23, ALBION. First shown on a map in 1861 (Lig.), but without name; first with the name in 1873 (H&W). But the name must have been in use much earlier.

*** Sweet School**
For A. L. Sweet, of Canada, and his descendants, who have held adjoining land since 1844. A former name for *Oaklawn School* (SUN PRAIRIE).

Syene, cor. [ˌsaɪ'in]
For the PO and station, and ultimately the prairie: a crossroad settlement in S sec. 11, FITCHBURG.

*** Syene Church**
For the *Syene* settlement, nearby: a Methodist Episcopal church, built between 1873 and 1877, in NW sec. 13, FITCHBURG. It was abandoned about 40 years ago.

*** Syene PO**
Apparently for the prairie. Est. July 21, 1864, disct. Aug. 3, 1885; re-est. Nov. 29, 1898, disct. Aug. 14, 1909. Kept in S sec. 11, FITCHBURG.

Syene Prairie
For the Biblical *Syene* (Heb. *Seveneh*)—*Ezek.* 29.10; 30.6. It was the southern frontier-post of Egypt; the name may therefore have been applied to the settlement which grew here in order to suggest its remoteness. The date and circumstances of the naming are unknown, however. It was settled at least as early as 1845, and included parts of secs. 10, 11, 12, 13, 14, and 15, FITCHBURG. (The Biblical name should be pronounced in three syllables; in this locality it has been reduced to two.)

Syene Road, The

Because it passed over Syene Prairie and thru Syene settlement: an early road running S from Madison into FITCHBURG. (Barton 1079)

Syene School

For *Syene Prairie;* it is in SE sec. 11, and was in existence before 1861, being the first school in FITCHBURG.

* **Syene Station**

For the prairie, and probably concurrently with the PO: a station of the C & NW railroad, established in SE sec. 11, FITCHBURG, about 1864; disct. before 1926.

* **Sylvan Lake**

For its location among woods; so named by A. O. Fox, owner of the land (Barry): a pond in NE sec. 35, FITCHBURG. The name is recorded only by Foote, 1890.

Table Bluff

An isolated, flat-topped bluff in NW sec. 28, BERRY; an old landmark, having several Indian mounds on top of it. So called as early as 1870, probably many years before.

Table Bluff School

For nearby *Table Bluff*. In SE sec. 29, BERRY.

Tamarack Swamp, The ['tæmə,ræk, 'tæm,ræk]

Descriptive: a sizable swamp (the only one of its kind in the locality) in S sec. 7, ALBION.

* **Tarpora Creek**

= *Nine Springs Cr.* (Possibly from Winnebago *tsha po ra,* "breast-bone.") This is one of the distorted Indian names used only on Greeley's map of 1854; it gained little if any currency.

* **Taychopera** (also, **Taychóberah,** etc.)

Winnebago, meaning *"The Four Lakes";* used also of the general region around them, and specifically of Madison (whence "Taychóberah Lake" was used of L. Mendota—Doty Let.). Chiefly used by the Indians; adopted, but never widely used, by the whites. First recorded 1837 by Featherstonehaugh; first in print 1840 (Doty Let.). For further details, see Cas. 1.

Taylor's Corners

For William R. Taylor (Governor of Wisconsin, 1874–6), who owned much of the adjoining land: the crossroads in the middle of sec. 9, COTTAGE GROVE. Taylor settled here 1848; he sold the land about 1903, but the name continues in use. A recent alternate name is *Kilian's Corners.*

Third Lake

An early name for *L. Monona.*

Thompson Creek

Because it runs from *Thompson Spring:* a local, unofficial name for *Badfish Cr.* near its source. (Ames)

Thompson School

For Gullick Thompson and his descendants, who held much adjoining land from before 1861 till after 1926. In NE sec. 3, CHRISTIANA. Now closed.

Thompson Spring

For Robert Thompson, who settled near it in 1842: a large spring in SE sec. 12, OREGON—now within the village of Oregon. (Still owned by Thompsons.)

*** Thompson's Tavern**

For Geo. P. Thompson, owner: a well-known early tavern on the old *Military Rd.*, at the present *Mackessey's Corners.* Thompson acquired it in about the late 1840's; it was razed in 1878. (Barton 1116)

*** Thornton's Marsh**

For C. C. G. Thornton, owner of the property from before 1890: a marsh W of the outlet of L. Mendota, in SE sec. 12, MADISON. Now a part of Tenney park, city of Madison (since before 1911).

*** Three Mile Creek**

Descriptive: a former name for *Marsh Creek.* So named only on Warren map, 1875, sheet 5.

Tiedeman School ['tidɪmən]

For John Tiedeman, who settled nearby before 1861, and his descendants, still on the adjoining land: in SE sec. 4, MIDDLETON.

*** Timber Island**

From its growth of trees: a small island in Hook Lake, sec. 29, DUNN. First shown and name recorded 1890 (Foote).

Tipple School

For David Tipple, of N. Y., and his family, who settled in SE sec. 7 in 1848 (Butt. 1201): the older name for *Oak Hill School;* still current.

Token Creek (formerly, **Tokun Creek, —— River, Tokunnee River**) ['tokən, 'tokn̩, 'tokŋ̍]

Origin uncertain, but most likely for Tokaunee, a mixed blood (Winnebago - Menomini) chief, who must have had something to do with the Winnebagoes who frequented the spot when the whites first came. The earliest maps (Cram 1839; Lapham 1846, 1848) spell it "Tokun," and Doty (Letter 1840) gives it as "Tokunnee"; not till 1853 (Lapham's map) does it appear in the modern spelling. Doty's spelling makes the connection with the chief, who is reported (De La Ronde's "Narrative," W.H.S. *Coll.* 7:359) to have had a small village of Winnebagoes near the present Mauston in 1837. Perhaps Tokaunee had had his followers at Token Creek before going to Mauston, and had left his name on the place, to be adopted (usually in abbreviated form) by the whites. He was not a prominent chief, which may well be why the origin of the name was soon forgotten. The modern spelling of the name is of a piece with the stories told to explain it—folk-etymologies, obviously.

Two such stories are offered: one, that the early white settlers found a "totem" here, which word was corrupted to "token"; the other, that the Indians were in the habit of dropping offertory "tokens" of tobacco for the spirits when they crossed the stream. Since "totem" is of Algonquian origin, and would not have been used by the native Winnebago, either of these explanations involves the white man, and neither could have preceded him. Thus, tho there is no record of chief Tokaunee's having been at this spot, to assume that he *was*, still gives the best explanation of the known forms of the name.

The stream now so called rises in SW BRISTOL and SE WINDSOR; thru the latter it flows SW into BURKE and W to join the upper Yahara R. on sec. 7. On early maps, however, the name was placed below this confluence (Cram), or on the (present) upper Yahara (Lapham 1855), whereas the former names of the Yahara were never placed above the Four Lakes. "Token Cr." thus formerly applied to the whole system of the upper Yahara with its tributaries; now it applies only to the chief eastern tributary. The first map to put Yahara above the lakes is Greeley's, 1854.

Token Creek, vill. (Also abbr. to "Token," see *Token PO.*)

For the creek. The village, tho never platted, was one of the earliest settled in this part of the country, and once quite important. It lies along the road between two crossroads, those in secs. 34, WINDSOR, and 3, BURKE.

Token Creek Hill

For the creek and villages, both at its foot: a conical hill in SW sec. 34, WINDSOR, rising to 1025 ft., or about 100 ft. above the surrounding area. Formerly *Big Hill;* also *Lawrence Hill.*

*** Token Creek PO**

For the village. Changed from Windsor PO, Feb. 25, 1869; changed to *Token PO*, June 25, 1883.

Token Creek School

For nearby *Token Creek.* In S sec. 34, WINDSOR.

Token Mill Pond

For Token Creek, both creek and village: a large pond or artificial lake resulting from the damming of Token Cr. in sec. 34, WINDSOR. It covers parts of secs. 34 and 35. (This is the upper millpond; there was a lower pond, long ago drained, in W sec. 4, BURKE.)

*** Token PO**

Name simplified, according to the PO policies of the time, from *Token Creek PO*, June 25, 1883; disct. Aug. 15, 1902.

Token Road

For Token Creek, which it crossed, and Token Creek village, which it passed thru: a road (est. 1844) from Madison to Portage (Eldred). From Madison it went NE to S-central BURKE, then N to Token Creek village, and slightly NW to Eagle Point (WINDSOR) whence it entered Columbia Co. Much of it has been abandoned, chiefly in WINDSOR. See *Ft. Winnebago Rd.*

Token Springs

The most important sources of Token Cr.: three springs in sec. 35, WINDSOR. (There are many other springs contributing to the creek.)

Tonyawatha, subd. [ˌtonja⊥ˈwɑθə, ˌtɑnjəˈwɑθə]

The locality about *Tonyawatha Springs.* A subdivision was platted out of this, by this name, in 1892, in NW sec. 17, BLOOMING GROVE.

Tonyawatha Springs

Apparently a pseudo-Indian name, made on the pattern of "Hiawatha" and such. The name does not appear to have existed before the establishment, in 1879, of the Tonyawatha

Spring Hotel, by a large spring "called Tonyawatha," which "gushes forth to gladden the sight and heal the infirm" (Butt. 929). The name, said to mean "healing waters," actually seems to be a pun on "tonic water," and thus to have been invented for commercial purposes. The area was first platted as *Tonyawatha*, then replatted in 1906 as Tonyawatha Springs; N sec. 17, BLOOMING GROVE.

Town Hall School

Because it is at the same crossroads with the town hall, between secs. 20 and 21, SUN PRAIRIE. Now closed.

*** Traveler's Home, The,** tree

Subjectively descriptive: a large bur oak in SE sec. 28, BURKE, so known to early settlers and travelers who rested under it. (Park 402)

Triangle School

Because it is in a triangular corner (extreme NW sec. 13, BURKE) formed by the junction of highway 19 and the section-line road. Now closed.

*** Troy**

Probably for Troy, N. Y.: a "paper town," platted June 24, 1837, by Eli Bates, Jr., of Milwaukee, in adjoining parts of NW sec. 13 and NE sec. 14, ALBION, surrounding the N end of *Rice L.* (which on the plat is called *Prairie Lake*). It came to nothing.

Truax Field ['tru,æks]

For Thomas LeRoy Truax of Madison, Lieutenant in the Army Air Force, who was killed in an airplane crash in 1941; the name was given shortly after: the Army Air Corps Technical Training School, covering parts of secs. 19, 20, 29, 30, 31, and 32, BURKE. (Barton)

*** Turkey Grove PO**

Probably for a grove which turkeys had frequented, tho no record has been found of such. Est. July 17, 1849, in SW sec. 31, PERRY; the name and location were changed to Moscow PO (Iowa Co.), May 3, 1854.

Turtle Lake

Origin uncertain; it may have some connection with Winnebago legends, in which turtles figure frequently; most likely, some settler named it so because it had many turtles in it (as it still has). A small lake, now virtually dried up, on the border of secs. 23 and 24, ALBION. It has always been shallow, with a variable amount of water; Lig. (1831) shows it as a swamp; H&W (1873) shows it as a lake, and first puts the name on it; USGS (1887–9) shows it as a lake larger than *Sweet L.*, which is usually shown as smaller. Recently, also *Lotus L.*

Turville's Bay ['tɝvɪlz, 'tɝvl̩z]

The inlet in the S shore of L. Monona formed by *Turville's Point;* SE sec. 25, MADISON.

Turville's Point (also, **Turville** ——)

For H. Turville, who settled here before 1861, and his descendants, who still own it: a point of land jutting into SW Lake Monona; NE sec. 5, MADISON.

Turville's Woods

Descriptive: *Turville's Point* was heavily wooded.

*** Twin Lakes**

Subjectively descriptive: two small ponds, close together but of unequal size, in SE sec. 26, FITCHBURG. They are shown on a map first by Foote (1890) and named only here.

Twin Valley School

Descriptive; it is below a considerable bluff which divides the upper valley of Black Earth Cr. between secs. 8 and 17, MIDDLETON. The school is in extreme SW sec. 8.

Tyvand School ['tɪvənd, 'taɪvənd]

For J. P. Tyvand and family, who from before 1890 owned the adjacent land in N and W, the school being across the road in NE, sec. 35, PERRY.

*** Underwood Hill**

For R. Underwood and his family, who settled in 1845 in sec. 3, OREGON: a former name for *Sholts's Hill.*

*** Union Canal**

A name for the portion of the Yahara R. uniting lakes Mendota and Monona, which was dredged and straightened in 1849. (Durrie's map, 1856.)

Union Valley

Origin uncertain, whether literally because this valley results from the union of two others, or is unified with Black Earth Valley, or metaphorically referring to the union of its inhabitants; the name seems to be disused except in the name of the school: a valley running SW in secs. 5, 6, and 7, CROSS PLAINS.

Union Valley School

For *Union Valley*, at whose mouth it is: in NW sec. 5, CROSS PLAINS. Now closed.

University Addition, subd. (MADISON)

For its being added to the original plat of Madison and joining this with the grounds of the University of Wisconsin. Platted 1839 by Frank Hudson for Simeon Mills, who probably named it. It included most of NW sec. 23. Since absorbed into the city of Madison.

University Addition, subd. (MAZOMANIE)

For its location in the N part of the village of Mazomanie, near the site of the projected Haskall University (Kittle 62–4): an addition, platted 1859, by William Robinson, one of the founders of the University, and therefore probably named by him.

University Bay

The bay in the S shore of L. Mendota between the University of Wisconsin frontage and Picnic Point; sec. 15, MADISON. So called at least as early as 1904 (USGS, Madison Quadr.), probably much earlier. Also, *Picnic Point Bay.*

University Creek (or, **University Bay Creek**)

Because it flows into *University Bay:* a small creek mostly in S sec. 15, MADISON.

University Heights, subd.

For its position on high ground adjoining the University of Wisconsin; platted 1893, in NW sec. 22, MADISON. Since absorbed into the city of Madison.

University Hill

As the site of the first University of Wisconsin buildings (Durrie, 1874; etc.). This name superseded *College Hill.*

University Park, subd.

For its position near University of Wisconsin grounds (and perhaps also bidding for patronage by University people); platted 1915, in NE sec. 20, MADISON.

Upper Mud Lake

By contrast with *Lower Mud L.* (i.e., Mud Lake proper) which is on the other side of L. Waubesa: the upper part of L. Waubesa (mostly in sec. 28, BLOOMING GROVE) separated from the main lake by the embankment and bridge of the CMSP railroad, built before 1854. How soon the name came into use is unknown; it is local only.

Upper Pond, The

A familiar local name for *Lake Marion*, which is the upper of two mill ponds. (Linc.)

Upper Sauk Road, The

Because it was N of the *Sauk Road:* a branch running from Springfield Corners thru Hyer's Corners, and Acorn, then NW thru ROXBURY. It was so called in the 1880's, probably before, and is still known to "old timers." (Brere.)

Utica, vill. ['jutɪkə]

For Utica, Oneida Co., N. Y., from which state, and perhaps also from which city, its early settlers came: a small settlement in S sec. 19, CHRISTIANA; begun 1845; never platted.

*** Utica PO**

For the village, at which it was kept. Est. July 28, 1848: disct. June 30, 1903.

Utica School

For nearby *Utica.* In SE sec. 19, CHRISTIANA.

Valdres Valley ['vɑldrɛs]

For the place of origin of most of its settlers (Valders, Norway): a small valley running from sec. 5, PERRY into sec. 32, BLUE MOUNDS. (Haugen)

Valley View School

For the view of Sugar R. Valley which may be had from it; in N sec. 29, VERONA.

*** Van Buren**

Probably for Martin Van Buren, just elected president: a "paper town," platted Dec. 19, 1836, by L. W. Weeks, F. W. Heading, and J. T. Haight. There is an error in specifying the position of the plat, however, which makes it seem to lie in three separate portions, one of them 6 miles from the rest. The other two, however, were to be in parts of adjoining secs. 36, RUTLAND, and 31, DUNKIRK; so the town was (as the plat shows) intended to be on both sides of Badfish Cr., on the S edge of the county. Early maps place it variously—some at the juncture of Dane, Green, and Rock counties; others elsewhere. It was never settled as such.

Vasen Hill ['vɑzn̩]

For H. Vasen, owner of the land from before 1911 till after 1935: a more recent name for *Bernards' Hill.*

Vasen Spring

For H. Vasen; see *Vasen Hill.* The more recent name for *Bernards' Spring.*

Vea Park, subd. ['viə]

For the Vea family (Martin and Fridtjof), industrialists of Stoughton (Barton); on L. Kegonsa, platted 1895 by Chas. T. Johnson, in SW sec. 30, PLEASANT SPRINGS.

* **Vee School** (also, **Wee** ——) [vi]

For an early Norwegian settler of this name; he was the father of L. L. Hulsether, and owned the land before him, i.e., some time between 1861 and 1873. It next became *Hulsether School,* and is now *Sunnyside School* (CHRISTIANA). (Onst.)

* **Veeder School** ['vidɚ]

For an early settler of this name (Onst.), of whom, however, no record has been found. A former name for *Rockney School,* now *Prairie Queen School.*

Veith's Park, subd. [vɑɪts, vits]

For Heio G. Veith, owner of the land; on L. Mendota, platted 1908; SE sec. 27, WESTPORT.

Vermont, town

For the New England state, from which some early settlers are said to have come (see *Vermont Valley,* however). It was formed Nov. 16, 1855, by separation from BLUE MOUNDS; it includes T7N,R6E.

Vermont Creek

For the town: it rises in sec. 36, flows N and NW thru secs. 25, 24, 14, 12, 11, and 2 into BLACK EARTH, where it joins Black Earth Cr. on sec. 26. The name is found on maps from 1906–7 (USGS) onward; it was in use much earlier, however, according to local informants. (Pauls., etc.) It is called *Black Earth Cr.* on platbooks from 1899, but this is now unofficial.

* **Vermont Settlement, The**

For the state of origin of the settlers: one of the early groups in RUTLAND, mostly in the SW part (Park 388); they gave it its name, and formed the only village in the town. Cf. *The Maine Settlement.*

Vermont Valley

As the valley of *Vermont Creek.* It runs N and S thru the E part of VERMONT. It was the site of an early Norwegian settlement, but from its position it may well be that it was first explored and named by some of the Vermonters who settled Black Earth, just N. If this is so, the name of the town may come from the valley, rather than the contrary.

Verona, town

For Verona, N. Y., the native place of George and William Vroman, who (with others) built in 1844 the Badger Mill, around which the earliest important settlement began. It was established Feb. 11, 1847, and includes T6N,R8E.

Verona, vill.

For the town and PO. The present village grew out of two distinct settlements: the first, begun about 1844, centered about Badger Mill and the Creek, in sec. 22. The second grew up about a mile N, around the crossroads in S sec. 15 (see *Verona Corners*). Building along the road between these soon made them a continuous settlement, tho only parts were platted (1886 forward). It was incorporated 1921.

Verona Cave

Because its entrance is in VERONA. Another name for *Richardson('s) Cave.* (Butt.)

Verona Corners

For the town: an early name for the crossroads in the N part of the present village of *Verona;* now used only by the older people.

Verona PO

For the town. Est. Feb. 23, 1847, and kept at the middle of sec. 22, by the Badger Mill. Before 1873 it was moved north to the "corners" in the S part of sec. 15. Both these sites are within the present village of Verona.

*** Verona Presbyterian Church**

The "First Presbyterian Church of Verona"; in NW sec. 20; center of a Scottish settlement (see *Scotch Lane*). It was built 1861, abandoned some time after 1911.

Viall's Spring [ˈvaɪəlz]

For Andrew Viall, of N. Y., owner of the property since 1859: a spring, obliterated when the Nakoma Country Club fairways were graded; the water still flows across the golf course, however, from NW sec. 33, thru S sec. 28, MADISON, and into L. Wingra. (Brown 7)

Vienna, town [ˌvaɪˈɛnə] also occasionally [ˌviˈɛnə]

For Vienna, Oneida Co., N. Y., from which state some of the early settlers came. It was est. by separation from WINDSOR, Mar. 2, 1849, at which time it included T8&9N,R9E. On Mar. 31 of the same year, however, with the separation from it of WESTPORT, it was reduced to its present area, T9N,R9E.

Vilas, the vill. ['vɑɪləs]

For the PO. This was the site of the original *Cottage Grove*, but when that name was taken, with the PO, to the present location, and the original village got the new Vilas PO, it came to be known as Vilas. It was never platted, however, and is now only a cross-road settlement.

*** Vilas Creek**

From *Vilas Woods*, thru which it partly flowed: a former local name for *Murphy's Creek*. (Brown)

*** Vilas PO**

For William F. Vilas (noted Madison lawyer, later Postmaster General and U. S. Senator), who owned considerable land about two miles N of the village of Vilas, where the PO was kept. *Cottage Grove* had been the name of both village and PO till 1882, when they were moved SE to their present location. The new name "Vilas" was chosen to replace "Cottage Grove." Est. Mar. 3, 1882; disct. Oct. 31, 1902. The name was perhaps suggested by C. E. Leslie, a prominent local musician (Cunn.).

Vilas School

For the village, just E of which it is; NW sec. 8, COTTAGE GROVE.

Vilas Spring

For L. B. Vilas, former owner of the property: a spring in NW sec. 34, MADISON; afterward called *Gay Spring*.

*** Vilas Woods**

For L. B. Vilas, former owner of this heavily wooded land (on parts of secs. 26, 27, and 34, MADISON); now a part of the University Arboretum and *Lake Forest*.

Village View School

Because it stands on rising ground, and commands a view of the village of *Deerfield*, in which it is.

Vincent School

For Samuel Vincent, of N. Y., and his descendants, who have owned nearby land since about 1858. The school is at the cross-roads between secs. 15 and 22, WINDSOR. Now closed.

Vinney O'Malley's Hill (also, **Vinney's Hill**) ['vɪni o'mælɪz]

For Vincent O'Malley, owner of the land since the 1890's: the highest hill in WESTPORT. Its summit is in N sec. 14.

Wa-che-etch-a, subd. [ˌwɔtʃɪ'ɛtʃə]

Probably a rendering of Winnebago *watci-ĕtca* [wɑtʃi'etʃɑ], meaning "at our home" (Lincn.), tho the plat records its meaning as "my home." On L. Waubesa; platted 1911, in E sec. 4, DUNN.

Waconia Park [ˌweˈkonɪə]

From the Dakota "wa-kó-ni-ya, n., a fountain or spring of water" (Riggs). Gay (1899) first shows this as merely "Waconia," perhaps the name of the spring there; it was platted as "Waconia Park" in 1909, in SW sec. 6, MADISON. The name is not native to this region, and was probably sought for in the dictionary, and respelled by its finder.

Wagner's School

For the Wagner family, whose members (H. and J. Wagner, etc.) have owned land in this neighborhood since before 1861. The school is in S sec. 24, SPRINGFIELD. Now closed.

* **Walbridge School** [ˈwɔlˌbrɪdʒ]

For J. G. Walbridge and his descendants, who held neighboring land (sec. 6, COTTAGE GROVE) from before 1861 till after 1890. A former name for *Seminary Springs School.*

Walters School

For H. Walters, who owned adjoining land from before 1926 till after 1931. The school is in SE sec. 24, CROSS PLAINS. Also called *Showers School.* Now closed.

Wangsness School [ˈwæŋzˌnɛs]

For S. L. Wangsness? He owned adjoining land in sec. 24 around 1911. The school is in E sec. 24, VIENNA.

Warner Creek

For Col. E. N. Warner, former owner from about 1910 of much land around it (Brown): a creek rising in sec. 19, MADISON, and flowing N into L. Mendota at Spring Harbor.

Washington Park, subd. (DUNN)

Named by William E. Colladay (who owned the land and had it platted) for the city of Washington, D. C., where he spent 26 years as a member of the U. S. Treasury Dept., and for which he had great affection; on L. Kegonsa, platted 1905; secs. 14 and 23, DUNN. (Coll.)

Washington Park, subd. (BLOOMING GROVE)

For its location along E. Washington Ave.; platted 1906, in the center of sec. 6; since absorbed into the city of Madison.

* **Waterbury School** [ˈwɑtɚˌbɛrɪ]

For J. R. Waterbury, of N. Y., who, in 1836, settled in SE sec.

18, DANE, close to where the school was later built, and remained till after 1873: a former name for *Willington School.*

Waterloo Creek

Probably for the nearby PO and settlement of Waterloo, Jefferson Co., thru which it flowed. The PO was established in 1844; the earliest found use of this name for the creek is in 1845 (Dane 117).

Since there were already at least 11 Waterloos in the U. S. by 1842, the name may have been brought by a specific settler, or may merely have followed the vogue. It may be significant, however, that many Prussians and English were among the early settlers. The creek rises N in BRISTOL, and flows SE thru YORK and MEDINA into Jefferson Co. and the Crawfish R. See also *Nauneesha R.*, the earliest name, and *Crawfish R.*

*** Waterloo Precinct**

For *Waterloo Cr.*, running thru it: an election precinct, est. July 9, 1845, and including the present MEDINA.

Waterman Springs

For the Waterman family (John, James, Horace, and E.) who settled in sec. 20, RUTLAND, in about the 1840's. The springs, mostly on their lands in the S part of the section, feed the Rutland Br. of Badfish Cr.

*** Watertown and Madison Railroad**

For the main towns on it: another name for the *Madison, Sun Prairie and Watertown RR.* (Lig. 1861)

*** Watzke's Lake** ['wɑtzkɪz]

For Franz Watzke, of Prussia, and his family, owners from 1867 till before 1899; a former name for *Schroeder's L.* (now *Kalscheur's L.*).

Waubesa Beach, subd. [ˌwɔ'bisə]

For *Lake Waubesa,* on which it is; platted 1907, in adjoining parts of secs. 5, 7, and 8, DUNN.

Waubesa Beach School

For the nearby *Waubesa Beach.* In NE sec. 6, DUNN. Now closed.

Waubesa Lake

= *Lake Waubesa*

Waubesa School

For the lake nearby. In SW sec. 35, BLOOMING GROVE.

Waucheeta, subd. [ˌwɔˈtʃitə]

Of unknown origin; an Indian or pseudo-Indian name. Platted 1935, on L. Waubesa; NE sec. 5, DUNN.

*** Waukoma Creek** [ˌwɔˈkomə]

For the village of Waucoma (note the spelling), platted 1845 on the creek (=*Badfish Cr.*) in the E half of sec. 6, PORTER, Rock Co., and E of the "Main St." of Cooksville. The word may be Ojibwa *wākomā*, meaning "crooked back," tho the allusion is unknown. The word "Waucoma" is used in Cooksville in the names of clubs and other organizations, but appears in the name of the creek only once (USGS map, 1887–9). Badfish Cr. was the name from 1830 or before, and is the only one today.

Waunakee, vill. [ˌwɔnəˈki]

Named by Louis Baker, G. C. Fish, and S. P. Mardin, who were planning the village in anticipation of the coming of the railroad; they chose this name from a list of possibilities submitted by Gen. Simeon Mills, the Madison pioneer, whose aid they had requested. Baraga gives Ojibwa "*Wánaki*, I inhabit a place in peace," and this may well have been Mills's source. The popularly accepted translations are, "pleasant land" or "pleasant earth," as if the end of the word were Ojibwa "Oki, earth" (cp. Okee, Milwaukee, etc.). The spelling was probably Mills's. When the shift of stress came about is unknown; it is certainly unusual, being progressive rather than regressive.

The first plat (1871) included part of NW sec. 8, WESTPORT; the village was incorporated 1893, and now includes part of sec. 5 also.

Waunakee PO

For the village. The name and location were changed from *Leicester PO*, Oct. 11, 1871.

*** Waurooshic River** (or **Creek**) (variant forms below)

Origin unknown; possibly Winnebago, meaning "chain, cord, string" (Lincn.). Used first by Cram, on his map of 1839, for the E Branch of Mounds Cr. Doty's Letter (1840) spells "Waurooshik," but his map (1844) changes it to "Wauroostic" (which is evidently an error). Lapham (1846 and 1848) follows Cram correctly. Chap. (1855) misspells "Wanrooshic."

*** Webber Spring**

For M. C. Webber, and his family, who owned the land from before 1873 till after 1911: a spring rising in NE sec. 8, PRIMROSE, flowing into the W Branch of Sugar R. (Moore)

*** Wee School**

Another spelling for Vee School.

Weinberg ['vaɪnbɛrç]

German, meaning "Vineyard-hill": a patch of land N of the highway and about half a mile E of Marxville in sec. 3, BERRY, where an early experiment in wine-growing was made. So called in the large local German settlement. The name is remembered, tho no longer much used. (Ketelb.)

Welch Hill

For Michael Welch and his family, who settled on it some time before 1873, and continued there till after 1904: a sharp hill in N sec. 15, OREGON. (Ames)

Wendt('s) Corners ['wɛnts]

For J. Wendt, who had held adjoining land from before 1911: the crossroad where secs. 1, 2, 11, and 12, YORK, meet.

West Bay, subd.

From its position on the west shore of *Monona Bay;* platted 1908, in NW sec. 26, MADISON; since absorbed into the city of Madison.

West Branch (of Sugar R.)

Because it is the more westerly of two branches. It rises in SE BLUE MOUNDS, and flows SE thru secs. 6, 8, 4, 9, 10, 14, and 13, SPRINGDALE, in the last of which it is joined by *Deer Cr.;* it continues SE thru MONTROSE, where it joins Sugar R. on sec. 28. The name was used by Lig. (1861), but other platbooks omit it or change it to *North Branch, Middle Branch,* or *South Branch,* sometimes contradictorily; USGS calls it West Branch; some highway maps, erroneously, *Little Sugar R.* See *Eastern Branch.*

West Cambridge, subd.

For its location W of Koshkonong Cr. and *Cambridge;* platted 1849, in NE sec. 12, CHRISTIANA.

West End Addition, subd.

For its location within the village of Cross Plains; platted 1923; N secs. 3 and 4, CROSS PLAINS.

*** Western Addition to Madison,** subd.

Descriptive: an early plat (1837), recorded severally by its part owners, John S. Nicholas (May 7), Cornelius G. Palmer (May 12), and Josiah A. Noonan (Sept. 18). It covered parts of secs. 22, 23, 26, and 27, MADISON.

West Koshkonong

The W part of Koshkonong Prairie or the settlement on it; see the next two entries.

West Koshkonong Church

For its position on *Koshkonong Prairie:* the first Norwegian Lutheran church dedicated in America (Dec. 19, 1844); sec. 14, PLEASANT SPRINGS. The original church was later replaced by the West Koshkonong and Western Koshkonong churches. (Barton 1036–7)

West Koshkonong School

For its location on Koshkonong Prairie. In SE sec. 11, PLEASANT SPRINGS.

West Lake Park, subd.

Because it is to the W of L. Mendota; platted 1906, in E sec. 12, MIDDLETON.

West Lawn, subd. (also, erron., **Westlawn**) ['wɛstˌlɔn]

For the "West Lawn Farm" of H. C. Adams, its site; platted 1903, including adjoining parts of secs. 21, 22, and 28, MADISON. Since absorbed into the city of Madison.

West Lawn Heights, subd.

For the neighboring *West Lawn,* above which it is; platted 1908, in SE sec. 21, MADISON; since absorbed into the city of Madison.

* **West Oregon PO**

For its position in OREGON, in the "Story" settlement; the first postmaster, Alva J. Story, owned land in NW sec. 19. Est. Feb. 7, 1877; disct. Jan. 15, 1879. See *Story PO.*

West Madison (MADISON)

The west side of the city of Madison: a local name.

* **West Madison** (MIDDLETON)

For its location with respect to the city of Madison: a "paper town," platted June 23, 1837, in the W half of the NW quarter of sec. 22, MIDDLETON. (Other accounts misplace it, one putting it on Mud L., another saying that the Speedway Road ran thru it.)

West Middleton, cor.

For the PO; this crossroads settlement, following the names of its PO's (which see), has been called "West Middleton" and "Barwig." Both names appear together on platbooks from Gay (1899) onward, tho only the former is still in use.

*** West Middleton PO**

For its position about 3 miles due W of the first *Middleton PO.* Est. May 13, 1858; disct. Feb. 27, 1894.

West Middleton School

For its location in N sec. 29, MIDDLETON.

Westmorland, subd. (sometimes, **Westmoreland**) [ˌwɛst'morlənd]

The name is a combination of more than one idea. When J. C. McKenna, Sr., one of the owners, was going to make the subdivision, his family decided together on this name. Several other plats had recently been made to the west of Madison; so Mrs. McKenna suggested that this would be "more land" to be platted in the "west." J. C. McKenna, Jr., says that since the land was treeless and rolling it suggested "moorland," which idea was incorporated. Also, it was euphonious and reminiscent of Westmoreland, England. (The pronunciation of the name here, however, does not correspond with the English pronunciation.) Platted 1926, in NW sec. 28, MADISON.

West Point, subd.

From its position on a point of land in the west part of L. Mendota; probably also reminiscent; platted 1906, in SE sec. 6, MADISON.

West Point Park, subd.

From its position: see *West Point;* platted 1928, in SE sec. 6, MADISON.

Westport, town

For Westport, County Mayo, Ireland, from which Michael O'Malley, first town Chairman, had come. The settlement was preponderantly Irish, and O'Malley had no difficulty in getting the name accepted. The town (T8N,R9E) was est. Mar. 31, 1849, by separation from VIENNA.

*** Westport PO**

For the town. Est. May 17, 1854. It was kept at various points in the settlement, in secs. 22, and 26, then on the border of 21 and 22. Disct. Jan. 7, 1876.

*** Westport Station**

For the town. This was the original name of the present *Mendota Station,* from the coming of the railroad (1871).

West Stoughton, subd.

For its position with respect to Stoughton; platted 1893, in NE sec. 6, DUNKIRK.

West Wingra, subd.

From its position with respect to L. Wingra and earlier plats using the name; platted 1916, in NW sec. 28, MADISON.

Whalen's Hill ['hwelənz]

For James Whalen (who settled on it before 1861) and his family: a current alternate name for *Helm's Hill.*

Wheeler Prairie

For Edwin Wheeler, who settled on it in 1843. It is mostly in NE DUNKIRK, but a small part is in adjoining ALBION. The Wheeler Prairie Cemetery (sec. 24, DUNKIRK) was named for it.

Wheeler Prairie School

For *Wheeler Prairie,* on which it is. NE sec. 14, DUNKIRK.

Whiskey Creek

Probably from some connection with the nearby corner (now *Four-X*), at which there was a tavern much frequented formerly. The name is local and unofficial but well known, and there may be a grain of truth in the story told to explain it—most of which, however, is probably fictional, an *ex post facto* elaboration. It is included as a good sample of its type:

An early settler named Hank Lawrence, having been drinking at the nearby crossroad (subsequently known as Whiskey Creek settlement), was returning E with a jug of whiskey. Instead of crossing the creek by the ford, he tried to make his horse cross on the foot-bridge, but the animal took fright and shied, and the jug fell into the stream. The cork came out, and as the whiskey gurgled away it seemed to say, "Good-good-good-good-." "Yes," replied Lawrence, "I know you're good, but I can't get you!"

The name is widely known locally, but has never appeared on a map, probably because the stream is nothing but the upper *Yahara R.*, above the entry of Token Creek, and particularly in the neighborhood of the bridge on highway K where WESTPORT, WINDSOR and BURKE come together, and where the incident of the whiskey jug is said to have occurred. (Ellis, Lake, Corc.)

Whiskey Creek, cor.

For the nearby creek: a crossroad settlement in SE sec. 1, WESTPORT, which formerly had a store, blacksmith shop, tavern, etc.; it has become known more recently as *Four-X.*

White Clay Spring

From the greyish marl mud surrounding it: a spring, next E of

Big Spring, in SW sec. 27, MADISON; now within the University Arboretum. It flows into L. Wingra. (Brown 7)

*** White Cross Spring**

For the White Cross bottling works, which utilized it: a spring formerly flowing into L. Wingra; NE sec. 28, MADISON. It was filled in.

Whiteley Hill

For Joseph Whiteley (or Whitely), who settled on it before 1861: a sharp hill over which the road goes between SE sec. 28 and NE sec. 33, OREGON.

White Oak Hill

For the many white oaks growing on it: a large hill, chiefly on secs. 26 and 27, FITCHBURG. (Barry)

*** White River**

Probably an error for (or abbreviation of) "White Water Creek," former name of *Koshkonong Cr.* This name appears only on the plats of Clinton and the North Part of Clinton, 1836.

White School

For the White family; Addison White, of Vermont, settled here in 1849. The school was built on his land near the crossroads in sec. 7, VERONA.

*** White Water and Madison Road**

For its termini: an early road, in existence in 1845 (Dane 135), which passed thru SE Dane Co., and *Buckeye* corner.

*** White Water Creek** (also, **Whitewater** ——)

Descriptive, ultimately; probably a translation from Winnebago: an early name for *Koshkonong Cr.* It is used first by Lorin Miller, surveyor, on the map in his field notebook (1833) on the present ALBION. There is no evidence to show whether he got the name directly from local Indians or trappers, or mistook this creek for the present Whitewater Cr. (Walworth Co.) which also flows into Rock River. The name is a not unusual one (cp. "Minnesota," which is so translated sometimes).

*** Whittlesey's Marsh** ['hwɪtḷsɪz]

For Thomas T. Whittlesey, who bought land including it (E sec. 1, MIDDLETON) in 1838.

Wide Spread, The ['waɪdˌsprɛd]

Descriptive: a widened part of the Yahara R. (secs. 10 and 11,

DUNN) officially called *Mud Lake.* Used as early as the 1870's. (Park 488)

Wilke School ['wɪlkɪ]

For Carl Wilke and his family, owners of the adjoining land since before 1890. In NE sec. 31, VIENNA. Now closed.

Willington School

A blend of the names Wilson (for Pres. Woodrow Wilson) and Washington (for Pres. Washington), with an orthographic *l* added. (Loper) The name came into use in about 1920. At the junction in SW sec. 17, DANE. Formerly *Waterbury School.*

Willow Park, subd. (MADISON)

Descriptive; along L. Mendota, within the city of Madison, platted 1895; SE sec. 12 and NE sec. 13. The "Willow Park Land Co." owned the land before the plat was made.

Willow Park, subd. (WESTPORT)

Descriptive (there were some large willows here); platted 1910, in NW sec. 27.

Willow Springs, subd.

Descriptive; platted 1904, within the city of Stoughton; S sec. 8, DUNKIRK.

Windflower Hill

Descriptive; many pasqueflowers (similar to the windflower) grow on it: an occasional name for *Vasen Hill.* (Fraut.)

Windsor, town ['wɪndzɚ, 'wɪnzɚ]

For Windsor, Vt., birthplace of Mrs. J. E. Carpenter, who, with Mrs. Wm. Whitney, another early settler, suggested the name. (Park 254) It was accepted at once (fall of 1846) and became official with the establishment of the town, Feb. 11, 1847. This included at first T8&9N, R9&10E; but with the separation of VIENNA and WESTPORT (1849) and BURKE (1852), WINDSOR was reduced to its present area: T9N,R10E.

Windsor, vill.

For the PO, and ultimately the town. A settlement grew up first about the crossroads between secs. 29 and 30, then when the CMSP railroad was built close by (1871) and *Windsor PO* was moved to W sec. 29, a plat was made including the part of NW sec. 29 between the railroad and the section line (1873), and called Windsor. (The original plat has been added to.)

Windsor PO

For the town. Est. July 22, 1847, and kept by the postmaster,

Nathan P. Spaulding, in the middle of sec. 34. The name was changed to *Token Creek PO*, Feb. 25, 1869, and it was moved south into the village of Token Creek. Another Windsor PO was soon established, however, and kept in SW sec. 28, by Samuel A. Warner, from about 1869–70; it was moved about 1871–2 to the settlement (platted as Windsor village in 1873) and was kept at the store of H. W. Grauman. (Lake)

Wingra, subd.

For the nearby *L. Wingra;* platted 1909, in NE sec. 28, MADISON.

Wingra Beach, subd.

For L. Wingra, on which it is; platted 1932, in adjoining parts of secs. 27 and 28, MADISON.

Wingra Creek

Because it carried the off-flow from *L. Wingra* (and therefore this name is applied today chiefly to the upper part of the creek, whereas the alternate name, *Murphy's Cr.*, is applied chiefly to the lower part). This creek is shown on maps from 1846 (Jervis, Lapham) onward, but without name; first appearance of the name is on Greeley's map of 1854. The course of the stream, tho not altogether clear, since it ran out of the swamp surrounding the lake, has been fixed since about 1910 or shortly after, when it was dredged. (Some maps of this time therefore mark it "canal.") It now makes an arc thru secs. 27, 26, and 25, MADISON, to enter L. Monona between Olin Park and Pleasure Park.

Wingra Lake

= *Lake Wingra.*

Wingra Heights, subd.

For its position on high land above L. Wingra; platted 1895, in NE sec. 27, MADISON. Since absorbed into the city of Madison.

Wingra Hill, subd.

For its position on a hill above L. Wingra; platted 1914, in SW sec. 28, MADISON. Since absorbed into the city of Madison.

Wingra Park, subd.

For the neighboring L. Wingra; platted 1892, on adjoining parts of secs. 22 and 27, MADISON. Since absorbed into the city of Madison. The name was apparently in use before the plat was recorded: Foote (1890) shows the area as platted, with this name.

Winnequah Point ['wɪnɪˌkwɔ]

A portmanteau word, formed from the words "Winnebago squaw" (Brown); given by Capt. Frank Barnes in 1870 when he opened a dance-hall and picnic-ground here. On the spot had formerly been a Winnebago village (see *Old Indian Garden*) and the Winnebago Indians continued to frequent it; another early name had been *Squaw Point.* Barnes followed the current fashion and manufactured an "Indian name" for his new property out of existing elements.

The explanations offered later show no knowledge of this derivation. It is said to mean "good water" (Butt.) and it is "corrected" to "Minnequah, as the Indians pronounced it" (Barton). The former is only half sense, since *winne* does mean "water" in Ojibwa; but *quah* does not mean "good." The second explanation offers the Dakota word for "water" instead of the Ojibwa, but leaves the second element unexplained. Nor do we know what "Indians pronounced it" so, since the only ones here were Winnebago. (The name *Winnebago* is not a word in the Winnebago language, but a name given by the Ojibwa. Thus *Winnequah* would be as foreign to a speaker of Winnebago as to a speaker of English.)

Winnequah Point, subd.

For *Winnequah Point;* platted 1941, in NE sec. 19, BLOOMING GROVE.

Wisconsin Heights

For their position along the Wisconsin R.: a range of prominent bluffs, stretching S by W from sec. 24 to 35, MAZOMANIE. See *Battle Ground.*

*** Wood's Point**

For Abraham Wood, its second white settler. He came early in 1837 and left in 1839, but the name remained in use till much later. Other names for this point and the adjoining land (secs. 19 and 20, BLOOMING GROVE) have been *Old Indian Garden, Squaw Point, Strawberry Point, Winnequah, Belle Isle;* it is now a part of *Monona Village.*

Woodland, subd.

Descriptive; within the city of Madison, platted 1927; SE sec. 31, BURKE.

Woodlawn, subd.

Descriptive, and probably also reminiscent; an addition to

South Madison, platted 1898, in sec. 26, MADISON; since absorbed into the city of Madison.

*** Wood's Addition** (to Dunkirk)

For Franklin Wood, its promoter: a "paper" addition to Dunkirk (itself a "paper town"), platted Dec. 8, 1836, including the E half of the SW quarter of sec. 28, in what became DUNKIRK.

*** Wood's Landing**

Named for its location at *Wood's Point:* a steamboat landing on L. Monona. Used in the 1870's and 1880's.

Woodward Grove, subd. (also, **Woodward's** ——, and formerly, **Woodard** ——) ['wʊdwərd, 'wʊdərd]

For William Woodward, of Indiana, owner: a piece of wooded land on L. Mendota, in NE sec. 35, WESTPORT. Woodward had it platted in 1892, but the name was probably in use long before, since he settled there in 1852. (Woodw.)

*** Wossemon River** (variant spellings below)

Apparently Algonquian, tho the particular language is doubtful; it may mean "his father" (Geary, Lincn.). This is a former name of the *East Br. of Pecatonica R.*, used on early maps: Chan. (1839) Wossemon; Lapham (1846, 1848) and Chap. (1855) Wassemon. Also, Doty Letter (1840) Woseemon.

Wright's Valley

For an early settler by this name, who from before 1861 till before 1873 owned the land on which it is: a small valley running down from the bluffs to the Wisconsin R., in N sec. 7, ROXBURY. (Derl.)

Wyldhaven, subd. ['waɪld,hevn̩]

A pun on the name of F. Wildhagen and his descendants, who owned the land from before 1873 until it was platted, 1935; probably also influenced by the nearby Fairhaven and Springhaven; on L. Monona, in adjoining parts of secs. 17 and 18, BLOOMING GROVE.

*** Wyseora Creek**

= *Burke Cr.* and the lower part of *Starkweather Cr.* This is one of the pseudo-Indian or distorted names used only on Greeley's map of 1854; it gained little if any currency. It has been explained as meaning "Cedar tree duck" in Winnebago (Brown MSS), but this is doubtful.

Yahara Park, subd.

For its location along the Yahara R.; platted 1909 by Louis J. Alsmo, in N sec. 14, DUNN.

Yahara River [jəˈharə, jɑˈhɑrə] (Corc.): [ˌjoˈhærə]

This is supposed to be from the Winnebago, meaning "catfish," the former (and still common) name of the stream. However, it is either a bad translation, or the actual Winnebago word (*Howix-ra* [hoˈwiçˌra]) has been greatly distorted in being "simplified" for speakers of English. Since the name is first used on Greeley's map of 1854, in common with the new Indian names of the Four Lakes and many other such names for lakes and streams never so called before or since, the form of the name may be laid at the door of Gov. L. J. Farwell, chief promoter, for whom Greeley's map was made. Tho Farwell had the aid of Lyman C. Draper of the State Historical Society in naming the Four Lakes, there is no evidence that Draper had anything to do with "Yahara" or the nonce names of Greeley's map.

"Yahara," with the names of the Four Lakes, was made official by a legislative act, passed Feb. 14, 1855, but its acceptance has been slow and never complete.

The stream rises in the extreme S border of Columbia Co., flows into Dane, passes thru WINDSOR, BURKE, and WESTPORT, thru the Four Lakes, then thru PLEASANT SPRINGS and DUNKIRK into Rock Co., to join the Rock R.

Other names for this stream have been *Coscahóenah R.*, *River of the Four Lakes*, and *Catfish Cr.;* the upper part is also called *Whiskey Cr.* and *Holum's Cr.*

Yellow Banks

Descriptive: the banks below which flows Marsh Creek; sec. 7, MAZOMANIE, Dane Co., and running on into Iowa Co. (Shown only by Warren, 1875, sheet 5.)

Yellowbanks Road

For the *Yellow Banks:* a road running along the Wisconsin R. in N MAZOMANIE, then turning S toward Mazomanie village. (Derl.)

York, town

Origin uncertain. It is said in county histories to have been named by English settlers, but since few English came before 1848, tho many settlers came from N. Y. state, then commonly called "York State," the latter is the more likely derivation.

Est. Mar. 11, 1848, by separation from SUN PRAIRIE; T9N, R12E.

York Center, vill.

For its location at the exact center of the town. Settlement began here about 1848, in which year the first store was built. It was never platted.

York Center School

For nearby *York Center.* In NW sec. 27, YORK.

*** York PO**

For the town. Est. Jan. 4, 1849; disct. Dec. 1, 1875. It was kept at York Center.

ADDENDA

Entries in brackets refer to the main list of Dane County place-names.

*** Bachelor's Grove**

An early name for the land covered with maple trees which shortly became *Farwell's Point.* The owner was unmarried at the time when the name was given. (Madison Daily Argus and Democrat, III, 205; Sept. 1, 1853)

Barney Lake

Another name for *Lake Barney.* (Derr Map Studio, *Atlas of Dane County Wisconsin,* 1954, p. 216)

Big Marsh (MADISON)

Descriptive. A marsh in the NE quarter of sec. 27, along the outlet of *Lake Wingra.*

[**Blue Mounds PO**]

In the 1890's the name was temporarily simplified by the PO Dept. to "Bluemound." (Brig.)

Bohn's Creek

Another name for *Elvers Creek,* which, above Elvers, runs across the land of J. F. and Ernst Bohn (H&W) in adjoining corners of secs. 27 and 34, VERMONT. The Bohns came in between 1861 and 1873.

Brickson School

A former name for *Buckeye School.* D. Brickson (or Brixon, as given on platbooks) held the adjoining land in the 1890's. (A. Duckert)

[**Buckeye School**]

It was formerly *Door Creek School,* and before that *Brickson School;* also *Pete Lee School.* (A. Duckert)

Buckeye Store

A store run by Mert Emerson at the junction of highway 12/18 and road MN, just W of *Door Creek* corner, from the early 1900's. It took its name from *Buckeye* corner, about $\frac{3}{4}$ mile E.

[* **Buckeye Tavern**]

It was on a section of road abandoned before 1890 (Foote), the remainder of which is present Schadel Rd. The name was extended to *Buckeye* corner, *Buckeye School,* etc.

Campbell Hill

A prominent hill chiefly in the NE quarter of sec. 33, COTTAGE GROVE, but also in adjoining secs. The land was owned from before 1861 (Lig.) by C. Campbell and his family. It has since been cut down. (A. Duckert)

Carry, The

Descriptive. A canoe portage across the narrow neck of *Picnic Point.* (Flint)

Clarkson, cor.

A crossroads settlement in the NE quarter of sec. 14, PLEASANT SPRINGS. It takes its name from *Clarkson PO,* which was close by.

[* **Dead Lake**]

The name perhaps referred to its lack of a clear outlet, since the lower end was marshy at that time. (*Wis. Mag. Hist.,* VI, 212)

[**Door Creek,** cor.]

Further relocation of roads places it (1968) at the intersection of highway 12/18 and road N, close to the middle of sec. 33.

* **Eagle's Nest**

Another name for *Eagle Heights,* current about 1900. (Flint)

Emerson's (Corner)

The intersection of highway 12/18 with road MN in sec. 33,

COTTAGE GROVE. Here Mert Emerson kept the *Buckeye Store.* The Emerson family were on the land here before 1911. (Cant.)

Five Points (MADISON)

Descriptive: another name for *Union Corners.* (Flint)

Hammersley('s) Pond

A pond in the SE quarter of sec. 30 and (mostly) the NE quarter of sec. 31, MADISON. It is on land owned before 1860 by the Hammersley family: in 1861 by W. Hammersley (Lig.). Hammersley Road, Madison, preserves the name.

Indian Pond

A small pond on *Picnic Point;* so called as late as 1900 because the Indians frequented it formerly to trap muskrats. (Flint)

Lake of the Evening

A supposed translation of the Indian name of *Lake Monona,* and used of that lake some time before 1947. (Lowry Axley, Savanna, Ga.)

Lake of the Morning

A supposed translation of the Indian name of *Lake Mendota,* and used of that lake some time before 1947. (Lowry Axley, Savanna, Ga.) This lake is more easterly, Lake Monona more westerly, in the Madison area.

Lieutenant Governor's Island

Humorously descriptive: a smaller island close to *Governor's Island.* This was another (and perhaps earlier) name for *Rocky Roost.* (Flint)

Marsh Corner

Descriptive: a marshy indentation in the shore of Lake Mendota at the E side of *University Bay.* So called about 1900. (Flint)

* **Marston's Woods**

A woods owned by F. T. Marston in adjoining parts of secs. 28, 29, 32, and 33, MADISON, in which *Nakoma* was platted. (G. H. Doane)

* **Pete Lee School**

A former local name for *Buckeye School.* It preceded *Door Creek School.* (A. Duckert)

* **Roderman's Woods** ['rɑd ɚmənz]

A former name for the woods platted (1924) as *Fuller's Woods.* It was named for John R. Roderman, who owned the land from the 1850's into the 1880's. Fuller acquired the property before 1890. (Lig., Foote)

*** Telegraph Road, The**

The road which became highway 14, between Oregon and Madison. So called because the first wire into Madison was strung along that route in 1849. The name was still in use in the early 1900's. (Dow)

Twelve Mile Hill

Descriptive: it is twelve miles from Madison. It is in the NW quarter of COTTAGE GROVE. (A. Duckert)

Union Corners

Descriptive: the point, now within the city of Madison, at which East Washington Ave. is joined by North, Milwaukee, and Winnebago Streets in the NE quarter of Sec. 6, T7N, R10E. (Flint) (Union St. is also only one block E.)

*** Winnebago Prairie**

A prairie which in 1840 stretched from about 4 miles N of Madison for over 30 miles in the direction of Poynette. The Winnebago Indians ranged over it. (A. W. Schorger; *Wis. Mag. Hist.*, VI, 212)

Appendixes

APPENDIX I: General Information

THE FEATURES NAMED

Place-names may be divided according to the types of features named, the first big division being into *natural* and *artificial* or man-made features. This holds well in most instances, tho not in all: woods, harbors, ponds, etc., are sometimes partly or wholly artificial. In this study, the majority has decided where individual features are to be placed; thus, even such things as drainage ditches or canals (always partly artificial) are listed with other water-courses. The explanations following the list of terms should prevent misunderstanding.

The names themselves are generally made up of two parts: the *generic*, which designates the feature, and the *specific*, which particularizes this one. Of these two, the generic may or may not be present, but the specific must always be; thus the specific part may be considered the essential name. For example, there are *Lake Mendota* (generic + specific) and *Mendota Village* (specific + generic); either of which may be called merely *Mendota* (the context making it clear which is meant), but neither of which would be a place-name at all without "Mendota."

One might go so far as to state that a place-name comes into being when the *specific* part is first used. Thus an observer might find *a spring*, and speak of it as such, but it has no name until he begins to call it *the Spring*, or (moving up the scale of specificity) *big spring*, *blue spring*, *Thompson's spring*, or the like.

It is interesting to note that tho some place-names may keep on or leave off the generic part, there are established customs in the matter. In general, the names for all natural features and most artificial ones express both parts; those for settlements often drop the generic part. The deciding factor seems to be the familiarity of the name, or how frequently it is used; context also plays a part, as was suggested above. Of the natural features only lakes and hills ever leave off the generic part altogether: *Mendota* (lake), *Round Top* (hill); rivers and creeks, when they leave it off, compensate by prefixing "the": *the Sugar*, *the Catfish*; hills, curiously, do a similar thing by prefixing "old": *Old Smoky*, *Old Pompey*. But all of these abbreviations are the exception, and would be used only informally. Wholly metaphorical names such as *The Devil's Chimney*, *The Traveler's Home*, may even add a literal generic term (rock, tree) unless context makes clear what features are meant.

On the other hand, the names of settlements, particularly the larger ones, do not usually keep the generic part, or keep it only in formal (e.g., legal) use. Thus one almost always hears *Vermont* (for the town), *Stoughton* (for the city), *Waunakee* (for the village), *West Lawn* (for the subdivision). Smaller settlements or their nuclei (corners, taverns, hotels, stores, etc.) come in occasionally for the same treatment.

If two or more similar features have the same specific part, of course, the generic becomes necessary for distinction: the *city* and the *town of Madison*; but when the generic term is omitted in such a situation, the more familiar feature is understood to be meant—in this case, the city. The generic part sometimes becomes combined with the specific, and so partly disguised: *Deansville*, *Jimtown*.

Generic terms are often converted into the specific part (or part of the specific part) of other names: *Marsh Creek*, *Wingra Hill* subdivision; this calls more often for the new generic part, to avoid confusion. An interesting example of such a development, resulting in a pleonasm, is seen in *Pheasant Branch Creek*. The specific part of a name may include a series of generic terms picked up thru a successive building of names: *Mendota Beach Heights* subdivision, named for *Mendota Beach* subdivision, named for its *beach* along *Lake Mendota*. This kind of development, in fact, is most frequent among real-estate names (specially discussed below).

Terms for Natural Features

The generic terms, both obsolete and current, found in names of natural features in Dane Co. may be sorted into groups as follows:
Water-courses: Spring, River, Creek, Branch (1), Branch (2), Brook, Run, Fork, Ditch, Canal;
Bodies of relatively still water: Lake, Pond, Spread, Slough, Lagoon;
Coastal features: Harbor, Beach, Bay, Point (1), Island;
Wet lands: Marsh, Swamp;
Relatively flat lands: Prairie, Plain;
Relatively low areas: Valley, Hollow, Hole, Pocket, Bottoms, Cave;
Relatively high areas: Mountain, Mount, Mound, Hill, Ridge, Bluff, Hogback, Heights, Knoll, Point (2), Bank;
Rocky features: Rock, Stone, Terrace;
Vegetational features: Tree, Grove, Woods.

The regional usage of these terms should be of interest. For the numbers of each, and the types of names, see Chart I.*

Spring. Always a natural source. A great many are unnamed; only those remarkable for size or for some other fact are named.

River and *Creek.* There is little competition between these terms any more, tho "river" was formerly more common than it is today: see *Black Earth Creek*, etc. Only two "rivers" now flow thru Dane County, the *Yahara* and the *Sugar*, tho branches of the *Pecatonica* rise in it. The common term is "creek" until the stream gains considerable size. There was, it is true, much more water everywhere in early days, but the more frequent use of "river" then proceeded, more likely, from the hopefulness or exuberance of the explorers and settlers with their grandiose schemes of waterpower and navigation.

Branch (1). This is the word in its common meaning: a small stream which joins with one or more others to form a larger stream: *North Branch*, *Oregon Branch*; it is fairly common.

Branch (2). This has the less common meaning of a stream flowing from a spring, and even (the South-Atlantic sense) of a whole stream, not a small upper tributary. So, at least, it seems in *Nine Springs Branch*, now *Nine Springs Creek*, and in *Pheasant Branch Creek.* The second of these shows by its pleonastic addition of "creek" that "creek" is unambiguous locally, whereas

* Page 226.

"branch" is not. In both of these names, "creek" has won the competition against "branch" (2), once by replacing it, once by overlaying it.

Brook and *Run.* These are both rare. The latter is a short stream issuing from a spring. The former is coming to mean, in this region, a stream stocked with trout. With conservationists active, it is likely to increase in frequency.

Fork. As usual, a confluence of approximately equal streams.

Ditch and *Canal.* Both mostly artificial; the latter is applied to the dredged and channeled part of a natural stream: *Union Canal.*

Lake and *Pond* seem fairly distinct in application. "Lake" was apparently used by preference, and of bodies of water of very different sizes. To be called a "pond," the body of water had to be very small, usually on the land of one man, unless it was a mill pond, in which case it could be larger than many "lakes," even covering several sections of land: *Krogh's Mill Pond.* Only one mill pond achieved an alternative lake-name: *Lake Marion.*

Some of the early "lakes" were extremely shallow, therefore muddy or with vegetation covering much of their surface, so much so that "mud lake" and "grass lake" seem almost generic terms. An example of a "lake" so called for promotional purposes is *Silver Lake* (alias *Mud Lake*). In general, the term "lake" is used with less restraint, beside "pond," than "river" is beside "creek." The area covered seems more decisive than the depth.

Spread, tho recorded only once in a Dane Co. name (*The Wide Spread*), is found elsewhere locally. I have seen Lake Koshkonong described as a spread of Rock River, for example. (NED *spread,* sb., sense 3, gives an example from 1824; surely there are U. S. place-names that record this sense earlier? DAE does not list it.)*

Slough or *Slew,* though it has various senses in Wisconsin, appears only three times in Dane Co. names, all for natural backwaters along the Wisconsin River. (DAE sense 1.)

Lagoon, on the other hand, is used of an artificial channel or opening within a subdivision bordering a lake or river, and having the twofold purpose of draining low land and making a way for small boats.

* DAE = *Dictionary of American English.*

Harbor is used of an artificially enlarged inlet in a lake shore where a creek or river enters. (Chiefly used in the names of subdivisions.)

Beach, *Bay*, and *Island* have their usual senses.

Point (1) is a piece of land projecting into a body of water—common since the 17th century.

A *Marsh* is usually less wet than a *Swamp*; the former also contains light growths (grass and reeds), whereas the latter may have trees.

Prairie is the common term for relatively level treeless land. Height is not a prime factor; some prairies are in river valleys others above the surrounding country. Size varies considerably. Prairies, when the settlers came upon them, contrasted distinctively with wooded areas, and were dry and flat enough (even when "rolling") to be easily cultivated.

Among the prairie names there are two of special note: *Mound Prairie*, as an abbreviation of *Nine Mound Prairie*, which was named in 1840; and *Mound Prairie* in what became the town of *Dunn*, so designated in 1833–4 by the Government Surveyor. This seems as if it may be a generic term; if so, it considerably antedates the entries in DAE (1867 and after) for a special region in Oregon called Mound Prairies. The mounds in Dane Co. are Indian tumuli, however, whereas those in Oregon are apparently natural.

Plain enters into only one name, *Cross Plains*, which seems half imported; if incidentally descriptive, the term was at least not an every-day one, and smells of the geography book.

Valley carries a sense of more size and dignity than *Hollow* tho it is not necessarily larger. Thus *German Valley* is smaller than *Dunlap's Hollow*. "Hollow," however, is used only in originally English-speaking communities: see *Hessian Hollow*. It is occasionally neutral, as in the example given, but more often used with a tinge of humor, particularly when given a final r-sound: "Brag Holler" for *Brag Hollow*.

Hole (NED sb., sense 1) and *Cave* have their usual meanings.

Bottoms (as in *Saw* (*Mill*) *Bottoms*) is interesting as a word not common here, but which, once established in this name, was borrowed by the Norwegian settlers who followed the first Americans: *Saga Batomen*.

Pocket seems to be still in the stage of metaphor here, appearing in the *Devil's Pocket* (alias the *Devil's Washbowl*); however, the sense agrees with that listed in DAE.

Mountain is applied to only one feature in the county, and then only on early maps at a time when that feature was known at a distance: *High Mountain* and *Smoky Mountains* for the present *Blue Mounds*.

Mound, however, for this feature and others in the area, natural hills more or less conical, is noteworthy. This is NED *mound*, sb.[3], sense 3b, for which the earliest quotation is 1810. The Blue Mounds were known long before this date, but whether actually by this name is uncertain. First use on a map is from 1829, but there must be letters and other documents in which they are so named earlier—probably before 1810. If so, this word is distinctly American. (DAE has nothing on it.)

Mount, as used here, is covered by the DAE treatment.

Hill is the most frequent geographical designation—which says more about the topography than the toponymy, since the term has the most varied applications. "Hills" are of all sizes and shapes here, and of course there are a great many more unnamed than named. Apparently, they achieve names when they are relatively prominent, of striking shape, or are in the way, forcing travelers to go around or labor over them. Interestingly enough, the hill or mound is the only geographic feature that seems locally to invite personification. This is indicated by a sort of nickname: *Old Pompey, Old Roundtop, Old Smoky.*

Ridge and *Hogback* have the usual senses, but the latter is not common, and apparently belongs to the past.

Bluff is common, meaning always the prominent hill standing above a creek or river valley.

Heights is common in real-estate names (see below) but otherwise almost unused. The one instance when it is not used of a subdivision is in *Wisconsin Heights*, referring to a particular part of the river bluffs where the decisive engagement of "the only war ever fought on Wisconsin soil" took place: *Battle Ground.* "Heights," in this name, is very likely of a piece with "battle" for the engagement referred to, and "war" for the whole uprising, the "Black Hawk War"—attempts to add consequence to rather small-scale things.

Knoll, similarly, is a consciously dignifying word.

Bank has its common use, however, as does *Point* (2), referring to a piece of high land noticeably rising above its surroundings.

Rock, *Stone*, and *Terrace* refer to natural features, the last being therefore an unusual usage (NED sense 2a is closest). The word is found only on an early map: *Rock Terrace*. Recent uses in real estate names suggest artificial leveling.

Grove and *Woods* refer to natural features, in Dane County names. A grove is usually a growth of trees contrasting as a mass with surrounding prairie lands. It may be quite small: *Cottage Grove*, or quite extensive: *Norway Grove*, but this contrast gives it clearer entity than "woods." The equal number of names in which the words appear is deceptive; *woods* is by far the more common word, used for any piece of land on which the trees have been left standing, and merely preceded by the owner's name when it needs to be referred to. But this is in the class with X's farm, X's barn or sand pit—private names essentially. The only woods names listed, therefore, are those which became widely enough known to have a more public character. In one name: *Sugar Bush Grove*, "grove" is added pleonastically—evidently because the synonymous sense of "bush" has been forgotten.

Tree. Only well-known landmarks were listed.

There are also a few names in which the generic part, being clearly figurative, cannot be considered a natural-feature term: *The Devil's Washbowl.*

Several foreign names introduce their own natural-feature designations: *Kalkberg*, *Weinberg*, *Kohlmanns Buckel* (German); *Bitsedalen*, *Nilsedalen*, *Saga Batomen* (Norwegian). These are used only by members of the foreign-language group. English equivalents are either translations or originals.

Terms for Artificial Features

The generic terms found in names of artificial features fall into the following groupings:

Settled areas: County, Town, Division of an Area, City, Village, Subdivision, "Paper Town," Settlement, Precinct;

Foci of small settlements: Post Office, Church, School, Tavern, House, Place, Stand, (Surname)'s, Corner(s), Points, Cross-road(s), Junction;

Transportation features: Railroad, Highway, Road, Lane, Bend, Crossing, Ford, Bridge, Landing, Station;

Miscellaneous small features: Camp, Field, Park, Forest, Farm, Breaking, Cut, Mine, Mill, Fort, Battleground.

For the numbers of each, and the types of names, see Chart II.*

County. There is no fixed size for a county in Wisconsin. *Dane County*, with thirty-five townships, is comparatively large.

Town. This is the abbreviation used orally and in informal writing thruout the county for "township" as a legislative unit, as distinct from the surveyor's unit. The first towns included many townships, but as population increased, new towns were separated out of the old, reducing the latter until today most towns are one township in size. The word "town," when it is mentioned, preceded the specific part of the name: *Town of Vermont.*

When it follows the specific part of a name, "-town" is not (as above) an abbreviation for "township," but means a settlement, usually a small one. It is about equivalent to "-ville," and the two used to be virtual alternatives: *Baertown* or *Baerville*; *Coffeetown* or *Coffeeville.* The latter, however, is more vigorous today: *Skellyville*, etc. Neither *town*, in this sense, nor *ville* stands separately in Dane Co. names.

Division of an area. This, of course, is not actually a generic term; it merely describes what the generic term would mean if there were one. Names in this group indicate the rather unspecific division of some larger area by prefixing a compass direction to the name of that area (a town, prairie, settlement, etc.): *East Blue Mounds*, *West Koshkonong.*

City and *Village.* These were given fixed legal significance in Wisconsin by the passage of general acts in 1883 and 1889. Their classification depends on population and (for villages) area. These exact senses of the terms are generally reserved for legal use, and the terms do not normally enter into place-names here. Thus the Village of Middleton is simply *Middleton* (unless it must be distinguished from the Town of Middleton).

Both terms also appear in non-legal senses, however. "City" was an element of several "paper town" names: *City of the First Lake*, etc.; otherwise it has scarcely been used at all—a striking

* Page 227.

contrast with the custom in Missouri and elsewhere. "Village" has always been the common term for a small settlement, but does not appear in the name itself unless this is a subdivision name: *Sunset Village, Pilgrim Village.* In Dane Co., then, both *city* and *village*, when made a regular part of a name, have always suggested self-consciousness or promotional assertiveness.

Subdivisions form a large class of legally regulated settlement features, whose use of terms is rather special.

A division has been made between those original plats of villages which bear the main village name, and additions to these original plats. The first have been listed as villages: *Waunakee*, etc., the second as subdivisions; since it is the first which are normally the nucleus of incorporation and furnish the village name, whereas the second remain suburbs (legislatively a part of the town) or become absorbed into the expanding village or city, and lose their legal separateness.

When a subdivision is platted it is called a "plat," "re-plat," "addition," or "subdivision." This, the generic part of the name, like "village," is omitted in normal use. Sometimes, however, such a term has had to be included as part of a name in our list, because to have omitted it would have been confusing: *Prairie Addition.*

Subdivision names freely use both natural- and artificial-feature terms, in various combinations: *Camp* Washington, Ruddy *Camp*; *Wood*lawn, Shore*wood*, Shore*wood* Hills, etc. The following list shows their frequency, the hyphens indicating whether an element is found suffixed, or prefixed, or both. (More elaborate combinations are not indicated.)

77 -park-
21 -heights
18 -addition
13 -lake-
12 -view
11 -hill(s)-
10 -wood(s)-
9 -beach
8 -side
7 -lawn
6 -land(s)-, -bay-, -camp-, -ridge-

5 -place
4 -court, -crest-, -garden(s)-
3 -edge-, fair-, -grove-
2 -estates, -forest-, glen-, -harbor, -haven, prospect-, -spring(s)-, summit-, -terrace, -village
1 -acres, brook-, -dale, -front, -green, -homes, homestead-, -hurst, -isle, -knoll, -plat, -point, shore-, -slope, -square.

Compass directions and the names of trees are also frequent elements, and there are many names having only one element which are not in the list above: *Nakoma*, *Brockton*, etc.

The specificity of these terms is blurred in real-estate names, many of which are used more for their suggestion than for designation. The vastly popular "park," applied to anything from thick natural groves to streets meagerly planted with saplings, comes to mean no more than that there are trees. Thus it ranges from being honestly descriptive to leaning heavily on imaginings of the future. Nor is "park" alone; the merest bulges in the landscape, if they rise somewhat above their surroundings, may be called "heights" or "hills." "Ridge," "beach" and "bay" are less abused, but "harbor" finds itself dignifying small, artificially enlarged inlets. Perhaps the grossest misapplication is in the name *Rosedale*, granted to a flat piece of farmland, conspicuously roseless.

All the real-estate names, of course, are by their nature attempts to be attractive. They hint at things desirable in a home-site—that it has a *view*, or better, a *prospect*; that it is near a *lake*, *springs*, etc.; or (a popular and salubrious suggestion), that it is on *heights*, *hills*, a *ridge*, *crest*, *summit*, *slope*, etc. Some names suggest homely comfort: *garden*, *grove*, *home*; others seek an air of prosperity: *acres*, *estates*; yet others use literary or exotic words hardly found here in non-real-estate names: *brook*, *dale*, *glen*, *green*, *haven*, *hurst*, *isle* (in a French name, *Belle Isle*), *knoll*, *terrace*; a few suggest privacy: *court*, *place*; and a recent one, pioneering community spirit: *Pilgrim Village*. All such things are, of course, more or less subjective and hard to define. But since the dates of all recorded plats are known, we may get some hint of the changing styles in these names.

Park, as by far the most frequent, is the least changing. Appearing first in 1889, it was used between 1890 and 1900 in 40 percent of the subdivision names; between 1900 and 1910 in 42 percent; since then in about 25 percent.

Heights, since 1890, has been in from 5 percent to 15 percent; *lake*, since 1855 (excluding its use in the names of "paper towns"), has been in from 2 percent to 11 percent; *view*, since 1890, has been in from 4 percent to 7½ percent; *hill(s)*, since 1894, has been in from 2 percent to 8 percent.

Other names have had more noticeable fluctuations of popularity. *Wood(s)*, used from 1898 onward, has considerably increased in frequency since 1920. *Beach*, from one example in 1896, had a sudden spurt between 1904 and 1910, with six examples, after which it fell off, the only others being in 1924, 1926 and 1932. *Camp* had a similar spurt, all six examples coming within twenty years: 1893, 1898, 1898, 1905, 1907, 1913; it has been unused now for over twenty years. Perhaps the summer cottager no longer likes to be threatened with "roughing it."

Lawn had a less pronounced vogue, but should perhaps be mentioned. The first example is quite early, 1869, then after almost thirty years it is taken up again: 1898, 1903, 1908, 1915, 1916, 1929.

Some words seem definitely more recent than others. *Crest* is first used in 1910, then has increased: 1928, 1938, 1945. One may note also *garden(s)*, 1918, 1925, 1926, 1928; *terrace*, 1928, 1928; *village*, 1938, 1939.

These words enter into various combinations chiefly to avoid repetition. For example, successive subdivisions are made near a lake and take its name as their first element. The second (or generic) element then becomes the distinguishing one (reversing the usual rule), and as several combinations follow, each must be different. A typical series begins with Lake Wingra, on which six subdivision-names are based: *Wingra Park*, 1892; *Wingra Heights*, 1895; *Wingra*, 1909; *Wingra Hill*, 1914; *West Wingra*, 1916; *Wingra Beach*, 1932. So also the series based on Lakes Mendota, Monona, etc. The newer words, *crest*, *garden(s)*, *terrace*, *village*, *knoll*, *slope*, *acres*, and so on, bring some novelty after series such as the one above have become stale thru repetition.

Another kind of vogue may be seen when the same sort of name flourishes for awhile. There was an "English" period when *Hillington*, 1917; *Marlboro Heights*, 1918; *Hillington Green*, 1921;

Westmorland, 1926, and others, were created. There appears to have been a "patriotic" period when *Camp Columbia*, 1893; *Camp Dewey*, 1898; *Washington Park*, and *Lincoln Park*, 1905, came into being.

The attempt to get an out-of-the-way touch is fairly recent: *Interlake*, 1911; *Linden Hill*, 1914; *Glenwood*, 1916; *Briar Hill*, 1918; *Belle Isle*, 1928; *Wyldhaven*, 1935. However, Indian names (not numerous) are pretty evenly distributed down the years.

"*Paper Towns*" have been listed separately as a special type of subdivision. They are those earliest of real-estate schemes in which all the "improvements" were made on paper, and the lots sold at a safe distance to hopeful settlers. Sometimes they were actually surveyed and staked out, and their names often found their way onto maps; always these "paper towns" were platted to look like real cities with streets, canals, water power, mills. But the plats did not necessarily correspond with reality. When the Four Lakes region was being opened up, a considerable number of these were projected. Their naming followed the usual patterns for early settlements, except that they favored the word "City": *City of the First Lake*, etc. Four names include it between 1836 and 1857; since then it has not been used. Only those are listed as "paper towns" which were never settled, or which existed only on paper for several years before any settlement was made.

Precinct. Precincts were set up for election purposes thruout the county before the establishment of the towns. They followed the same course as was later taken by the towns, shrinking in area as increase of population led to their being divided. Thus the towns, as they were established, often took their names from the precinct which they superseded.

Post Office. It is no exaggeration to say that, in early days, this term referred more to an activity than to a place. The appurtenances of a PO were at least a box with pigeon-holes, facilities for handling letters and the sale of stamps, and a duly constituted postmaster or postmistress. But the kind of building in which the PO was "kept" did not matter. As often as not, it was merely a corner of a farm living-room. Abe Lincoln and his hat were a not unusual "post office." Thus the PO was easily moved, a rather unstable name-feature; if the postmaster moved, or if he were replaced, this usually meant a relocation of the PO. As rural routes have killed off the smaller PO's, however, the name has come more and more to include the building or rooms along with the activity.

PO names have always been subject to certain controls. No newly proposed name could be accepted if the same or a very similar one was already in existence in that state or a nearby state. Since the person asking for the PO was often unable to guard against such clashes, many a name was rejected. As the number of PO's increased, the choice of a new name became difficult; a village might not even be able to get a PO by its own name (see *Floyd PO*). Long lists were submitted to the PO Department, from which they would make a choice (see *Hamlin PO*, *Acorn PO*, etc.); on occasion, the lengthiness of this process became a source of exasperation (see *Peculiar PO*). The PO Dept. also changed the form of some names (see below, under Orthography).

Church. Much like "school" (see next), this generally involved a building. A congregation often was organized for many years, holding meetings in homes, schoolhouses, or other buildings, before it built a church. Most churches thus came after a settlement was well established, and merely took the settlement's name. For this reason, only those churches have been listed which formed the nuclei of settlements, or whose names spread to other features.

School. By contrast with the PO, this has always included the building in its reference. In early days they sometimes "kept school" in existing buildings, but a school name does not appear to have been bestowed until there was a building specifically for school. Schools were moved too (see *Howarth's School*), but not as often as PO's.

School names are of at least two kinds: those spontaneously given, and those given to satisfy the Wisconsin statute of 1916 requiring that all rural schools be named. The former are pretty closely connected with local history and local people: *Tipple School*, *Stone College*; the latter are often regrettably vapid: *Oak Hill School*, *Maple Grove School* (four of these!). (*The Frog Pond School* was recently renamed *Oak Lawn School*, apparently in the fancy that this savorless transformation was an improvement.)

Tavern, *House*, *Place*, and *Stand*. These seem to have had about the same meaning, since, in early settlements of which such establishments formed a nucleus, they often combined the functions of sheltering, feeding, supplying, and quenching the thirst of travelers. *Tavern* was probably the most common of these terms, *stand* the least. They so often went by the name of the man who kept them, that some names of this kind exist only as abbreviations: *Haney's*, *Brigham's*, etc. These must be listed separately—(*Surname*)*'s*—since it is not certain which of the possible generic terms has been omitted or is "understood."

Corner(s). This refers usually to any junction or intersection of roads at which are one or more buildings—a farmhouse, school, town hall, store, etc. Unless there is some settlement, it does not achieve a name. The plural form is decidedly more common. An occasional alternative is *store*, but the generic term may be omitted altogether: *Nora Corners*, *Nora Store*, or just *Nora*.

Points. Only in one name: *Five Points*. Exactly what the semantics of this may be is uncertain. Does it refer to the tapering lands between the five roads that meet, or to the roads themselves, pointing five ways like the points of a star? Or is the point at which the roads meet common to all five? The dictionaries (NED, DAE) give no help on this; yet it is certainly not an unusual term outside of the county. Since *Five Corners* is an alternate form, the first of the above explanations seems the most acceptable. Compare the *Four-X*.

Crossroads and *Junction*. These have their usual senses.

Railroad. These names have apparently from the start been composed of the names of towns or regions which the railroad served or was intended to serve. This generic term is sometimes abbreviated: *The Madison Branch Road*, *The Milwaukee Road*.

Highway. Only one example, since the present highways, known only by numbers, have not been included in the list. *The Aztalan Highway* has consciously a touch of the grand about it, for the settlers were proud of their roads, a sign of progressiveness.

Road. This, nonetheless, was the usual term, and still is. Most names of roads, like those of railroads, mention places which they connect; but others are descriptive: *The Military Road*; and two early ones bore the names of settlers whose places they passed: *The Haney Road*, *The Daniel Baxter Road*.

Lane is generally applied to any small or private road, e.g., one leading from the public road to a farm. The sense of privacy and narrowness appears to carry over when this word is used in names, tho it actually refers to part of a public road, the first word in the name indicating the special character of that part: *Scotch Lane*, *Maple Lane*.

Bend has its usual sense.

Crossing and *Ford* appear to mean the same, tho the former is the usual word, in early use, and may refer to crossing by boat. The latter is later, perhaps a literary use, chiefly limited to one author: see *Koshkonong Ford*. Crossings were at first forded by the Indians, then the whites; finally, the main ones were bridged.

Bridge thus became an important feature, and might combine with, or supersede, the word "crossing": *Black Bridge* (*Crossing*).

Landing is the usual word for a place where boats, particularly steamboats, took on and discharged their loads. One humorous name makes this a "station": *Angleworm Station*—apparently in imitation of the railroads.

Station. The number listed is small because the word has most often become part of the specific part of names of village plats or of PO's: *Mount Horeb Station*, vill.; *Middleton Station PO*. Only when a railroad station existed apart is the name listed separately. The railroads were at first of such importance to the settlers that the establishment of a station usually led to the platting of a village beside or around it. "Station" might then be included in the village's name, to advertise it, or to distinguish it: *Paoli* and *Paoli Station*. But this did not find favor for long. Perhaps such a name implied a settlement with a station and little else. Whatever the reason, "station" was soon dropped from some names. PO names were sometimes involved in this simplification: *Basco*, *Dane*. Occasionally a mere loading platform would be dignified locally with this word: *Beanville Station*.

Camp. This is found in the military sense: *Camp Randall*, and the nonmilitary: *Camp Gallistela*. Real-estate names have it only in the latter. It is usually the first element of the name. *Old Indian Garden* and *The Indian Village* have been counted with these; tho they do not contain the word, they were essentially camps.

Field has the military sense: *Truax Field*.

Park. Urban parks have not been listed. Names listed in this group (e.g., *Stewart Park*) have the word in its usual sense, not as it is used in the names of subdivisions (see above).

Forest. This is not used locally of natural growths of trees, except in real-estate names. The *Forest of Fame* is an artificial park.

Farm has its usual sense, but of course very few farm names have been mentioned, except as they became transferred to settlements, subdivisions, etc. The only farm names listed are those so well known as to be almost public property: *The '76 Farm.* The word "farm" is not regularly used in all such names, however: *Montjoy.*

Breaking. Recorded in only one name, but it was surely in common use. DAE, sense 1, has a first quotation from 1860, but it was in use here before that: *The Long Breaking.*

Cut. Found only in *The Rock Cut*; in the usual sense.

Mine. This word was used hereabouts regularly, but the synonymous "diggings" and "lead" were probably as widespread. All three terms were applied to Ebenezer Brigham's lead mines at Blue Mounds: *The Brigham Lead, Brigham's Diggings, Brigham's Mines.* All such names are counted together.

Mill. Mills are listed only when their names had importance in their locality, when they spread, etc.: *Badger Mills.*

Fort. The only example is *Blue Mounds Fort,* a small early fortification.

Battleground. Several early maps so indicate the site of the defeat of Chief Black Hawk in the "war" of 1832.

THE TYPES OF NAMES

Place-names may be classified in a variety of ways, depending on the degree of detail desired, and the direction of the investigation. Not every possible classification has been made here; it seemed preferable to let the names sort themselves out, so to speak. In other words, in the charts which follow, and which indicate the numbers found of each type of name, classes have been set up for only the more striking types. As has been suggested, the reader may go further if he wishes; the materials are arranged for easy handling.

The classes which appeared to be noteworthy in this region, then, are as follows:

1. For an important non-local person
2. For a local person—owner, neighbor, etc.
3. For a distant place
4. For a nearby place or establishment

5. Descriptive and locational
6. Subjectively descriptive
7. Inspirational and symbolic
8. Anecdotic, ironic, humorous
9. Uncertain and unknown
10. Punning and blending
11. Pseudo-Indian
12. Indian
13. French
14. German
15. Norwegian

Numbers 10 to 15, being concerned with linguistic matters, usually overlap with some other classification. Consequently some names are classified doubly (or even multiply): *Kohlmann's Buckel* as both anecdotic and German, *Winnequah* as a blend and pseudo-Indian. (These multiple entries explain the disagreement in the totals of the horizontal and vertical columns of the charts.)

Lest the terms, as here used, be misunderstood, they had better be explained to begin with.

For an important non-local person. This would include all honorific names: *Camp Dewey*, the town of *Burke*.

For a local person. With these, the importance of the person is not a prime factor. A former governor of the state is remembered in *Taylor's Corners*, but chiefly as a local landowner. This, *Hippe's Hill*, and a great many others make up one of the largest classes.

For a distant place. Names of this class include only the first of a series, often. For example, the town of *Albion* was named for Albion, Orleans Co., N. Y. The many local names here that used the word "Albion" subsequently to the naming of the town, are considered as being named for it, rather than for the New York place, and therefore belong in the next class.

Sometimes a name has been used so often elsewhere that its appearance here is less likely the result of invention than of recollection, perhaps unconscious, and perhaps of no single or specific place. Examples are *Riverside*, *West Pcint*, etc.; such names have been called "reminiscent."

For a nearby place. Here belong *Albion*, vill., *Albion Creek*, *Albion Marsh*, *Albion PO*, *Albion Prairie*, etc., all based on the town of *Albion*, Dane Co., Wis. This is easily the largest group.

Descriptive and locational. This implies an objective description: *Mud Lake, Prairie View School*; it should be contrasted with the next class. By locational is meant the use of a direction for identification: *East Branch* of Sugar River, *West End Addition*, etc. (On the other hand, *North Bristol* and *South Madison* belong in group 4.)

Subjectively descriptive. This includes those names in which some personal or individual observation or judgment enters, with which another individual might not concur. *Silver Lake*, used of the same body of water as the Mud Lake just mentioned, would be an example; or *Pleasant View School*, as contrasted with Prairie View School, just mentioned.

Inspirational and symbolic. These names are such as *Liberty Prairie* or *Hope*. Their subjective quality makes them analogous to those of group 6, except that these are not descriptive; they simply make reference to some idea or ideal which the namer wishes to express in connection with this place.

Anecdotic, ironic, humorous. These similar, and often overlapping types of names have been grouped together. Even so, they form a small class. It might be mentioned that the anecdotes told to explain names are by their nature suspect; much like folk-etymologies, they are often *ex post facto* concoctions, attempts to rationalize the name. Examples: *Whiskey Creek*, *Poverty Hollow*, *Hundred-Mile Grove*.

Uncertain and unknown. Here are included names lacking even the basis for a good presumption. If the uncertainty was but slight, another classification was used.

Punning and blending. These names overlap in classification with other types. For example, *Wyldhaven* must be listed in three places: it appears to be made partly from the name of former owners (class 2), the "Wildhagen" family, blended with part of the name of two early neighboring subdivisions (class 4), "Fairhaven" and "Springhaven." The first syllable of the family name, tho respelled "Wyld," is still intended to suggest a rural spot (a pun), and "haven" is probably intended to be figurative. See also, *Willington School*, *Glen Oak Hills*, *Winnequah*, etc. This class is small, but interesting enough to deserve notice.

Pseudo-Indian. This has only one clear example: *Tonyawatha* (also a punning name); Winnequah might possibly go here too, tho at least the parts of this blend are ultimately Indian. Since these are not genuine Indian names, however, it seemed best to set up a class, even tho very small.

Indian. These are comparatively few in Dane Co. Winnebago, Sauk, Fox, Prairie Potawatomi, and other Indian groups dwelt in or passed thru the county in early days, but relatively few of their names remained. Only those names are listed which found their way onto maps or into some degree of usage by the whites. Of these, a good many are found only once (e.g., *Eshunikede Lake, Oskaw River*), and others only a few times: *Waurooshic River, Peena Creek*. Some were translated, and therefore do not count as Indian names: *Catfish Creek*, the *Four Lakes*. Even of those adopted, the indigenous: *Lake Wingra, Lake Koshkonong*, etc., are fewer than the imported: *Lake Mendota, Waunakee, Minniwakan Spring*, etc. Some appear to have been chosen "for euphony" rather than for appropriateness of meaning.

In dealing linguistically with Indian names, the writer has depended heavily upon the authority of the Rev. James A. Geary (for Algonquian) and the Rev. Benjamin Stucki (for Winnebago). There is a dearth of reference works on the Indian languages of this region; without the generous help of these two consultants, therefore, the handling of the Indian names would have been sadly inadequate. The historical part of the treatments, however, is my own.

French, German, Norwegian. Names such as *Belleville* are not included here; the name is ultimately French, but the Dane Co. village was named for a town in Canada (class 3). Neither do such names as *German Valley* or *Norway Grove* come into this class; but *Weinberg* and *Nilsedalen* do.

LINGUISTIC ASPECTS OF THE NAMES

To attempt a very detailed linguistic analysis of so few names from an area set off so arbitrarily would seem of doubtful value. Generalizations would have to be hamstrung with qualifications; it therefore seems far better to postpone such conclusions until a much larger local area than Dane Co. has been studied. Some notes may be made at once, however, to point the way for the future.

Etymology. As has been said, this is not a major question in these place-names. Most of them are composed of obvious elements; the foreign ones are few. Remarks have already been made on *punning* and *blending* names. Only one example was found of *back-formation: Morland Terrace.* No names appear to be wholly formed by *folk-etymology*, but this process does enter in occasionally: *Brackenwagen's, Sow Bottom, Token Creek, Hoboken*; possibly in the variant *Bearville.*

The writer was once present at the creation of a folk-etymology: an old informant of foreign extraction, when asked the meaning of *Eagle Point*, took the first word to be "equal" (had he heard the dialectal "ekal"?) and suggested that perhaps the *point* was an *equal* distance from Milwaukee and Portage!

Far more important than etymology is provenience, and this is analyzed by means of the charts below.

Lexical and *dialectal* matters are dealt with in the discussion of the names of features—the generic terms—above. See especially the remarks under *branch* (2), *spread, hollow, mound, points*, and *breaking.* For loan-words into foreign languages noted in Dane Co., see *Die Sittlament, Saga Batomen*; for a foreign name gaining some currency among speakers of English, see *Halunkenburg.*

Orthographical factors are not particularly striking, but one might note the Englishings of Indian names, particularly *Mazomanie, Waubesa, Token.* The majority of Indian names are spelt with a final *-a*, representing a variety of sounds: [i, ɪ, ɛ, ɑ, ɔ]. (For details, see Cas. 1; note also *Nauneesha*). The sound [ɔ] is spelt variously with *-a, -aw, -uah* when final: *Nauneesha, Oskaw, Winnequah*; and *-a, -au* when internal: *Wa-che-etch-a, Waunakee.* Initial *wa-* may represent [ɔ, e]: *Wa-che-etch-a, Waconia.*

A few names composed of two elements have combined into one, but none have gone the other way. This was aided by official action of the PO Dept., which, in the 1880's and particularly in 1894, set out to simplify PO names. See *Springdale, Lakeview*; and for other types of simplification, *Martinville PO* and *Token PO.*

Phonological factors worth mention are somewhat more numerous, tho in general, Dane Co. place-names show little variance from norms of middle-western pronunciation. One finds *creek* pronounced informally with the vowel [ɪ], formally with [i]; an unstressed *lm* group tends to acquire anaptyctic [ə]: *Helm's Hill*; [ng] easily becomes [ŋg]: *Lake Wingra.*

In a few names: *Paoli, Vienna* (and there are others in neighboring counties), the pronunciation of the *i* has been altered from [ɪ] to [aɪ]. The history of this is not clear, but the alteration probably took place in the East and was brought here. There is some tendency to revert to original [ɪ].

Indian names, particularly those whose last syllable is a consonant (or consonant-group) *-a*, almost all take the main stress on the penult: *Kegonsa, Mendota, Monona, Nakoma, Waubesa, Waucheeta, Waukoma, Wingra, Yahara.* Even names with secondary stress do this: *Mazomanie, Tonyawatha, Wa-che-etch-a.* Exceptions are *Koshkonong, Nauneesha, Waunakee.* This tendency seems to have no regular relation to the stress of the Indian source-words, which has in many cases been altered. The most striking instance is *Waunakee*, in which the last-syllable stress is not normal either to the Ojibwa word or to English pronunciation.

Two more names stressed in an unusual way are *Springdale* and *Westmorland.*

Since most foreign names are either limited in use to the group that gave them, or are translated, the foreign sounds have little effect in English. The exception comes with surnames, which must be used by both English- and non-English-speakers. These acquire variant forms very easily, as English-speakers either do not produce the foreign sounds, or else give the names spelling-pronunciations. Some interesting results may be seen in *Danz School, Helland Spring, Nordness Corners, Erbe School, Ruste School*, etc. There seems to be little uniformity in the degree of Anglicization, or in the direction this process takes—for example, in the retention or loss of final syllabic *-e*. But what is needed here is an independent study of local personal names; place-names alone give too little evidence.

Structure. This is discussed above with the explanation of the terms *generic* and *specific.* The generic term, when it is used, comes second, except in the formal phrases *city of—, village of—*, etc., and in the names containing the words *mount, camp, lake*, and *lagoon*, in which usage is divided. *Mount* always precedes the specific part; *camp* usually precedes; *lake* precedes in Indian names, but follows in others: usually *Lake Monona, Lake Koshkonong*; but *Indian Lake, Goose Lake, Sweet Lake*, etc. The influence behind this is French word-order, of course, in which the terms *mont, camp, lac*, and others would come first. An interesting case in point is in the names of the channels in the Belle Isle subdivision: pseudo-French *Lagoon du Nord* and *Lagoon du Sud*, along with *Sumac Lagoon*!

Note also remarks on *-ville* and *-town*, under discussion of the generic term *Town*.

LOCAL SPREADING OF NAMES

From the list of names have been taken all those in which the specific part has spread to one or more other features than the one which first bore it. One hundred and sixty-eight such groups were found, containing from 2 to 12 members. The sequence of spread is chronological, not directional. Thus, such a sequence as village—PO—creek means not that the village name, transferred to the PO, was thence transferred to the creek, but that the PO got the name later than the village, and the creek later than the PO. When there was any serious doubt of this order, the names were omitted.

The groups are listed in the various sequence-patterns in which they appeared, and in descending frequency. Similar features have been lumped: Ridges and bluffs with hills, corners with villages, hollows with valleys, rivers with creeks, groves with woods.

Village—PO (9 of these)
“ — “ —Valley
“ — “ —Creek
“ — “ —Town
“ — “ —Village
“ — “ —Precinct
“ — “ —Subd. (2 of these)
“ — “ —School (4 of these)
“ — “ —Hill?—Creek?—School
“ —School (12 of these)
“ —Creek
“ —Station—PO
“ —Vill. —Town—PO
“ —Paper Town—PO —Subd.—Town—2 Subd.—Creek
} 37

PO—Village (5 of these)
“ — “ —School (6 of these)
“ — “ —Village
“ — “ — “ —School
“ —Town
“ — “ —Village—School
“ — “ — “ —Village—School
“ —School (2 of these)
“ — “ —Village (2 of these)
“ —Station— “
“ —Plains — “ —Precinct—Town—Prairie
} 22

Town—PO (2 of these)
" —" —Vill.
" —" — " —School
" —" — " —Vill. —Cave
" —" — " —Road—School—Creek
" —" — " —PO —Vill. — " —School
" —" —Station
" —" —School
" —" —PO—Vill.—Vill.
" —Vill.
" — " —PO
" — " — " —Creek —PO
" — " — " —Hill —School
" — " — " —Prairie—Marsh—School
" — " —Vill.—PO —PO
" —School (2 of these)
" —Creek—Valley
" —Paper Town—PO—Vill.—School

} 20

Creek—Vill.
" — " —PO —Spring
" — " — " —Vill.—School
" — " —Road—PO —Springs?—Hill?—Lake—School
" —Valley—School
" — " —Vill.
" —PO — "
" —" — " —School
" —School (2 of these)
" —Hill
" —Subd.
" —Ford —School
" —Prairie—Hill—School

} 14

Lake—Creek
" — " —Prairie—Settl.—Ford—Div. Area—School
" — " —6 Subd.
" —Hill
" —School (4 of these)
" —PO —School
" —Subd. —Subd.—Station—Vill.—School
" —Point —PO —Vill. —Hill —2 Station—School—4 Subd.
" — " —Camp—2 Subd.—Ford—Bay —4 Subd.—Vill.
" —Station—Subd.—2 Schools

} 13

Hill—School (7 of these)
" —Vill. —School
" —2 Subd.—Bay —Subd.
" —PO —Vill.—School—Vill.
" —Settl. —Fort—Road —2 Creeks—PO

} 11

Valley—School (6 of these)
" — " —Creek } 8
" —Hill —School

Spring—Subd. (2 of these)
" —2 Subd.
" —Creek (2 of these)
" — " —Marsh?—Prairie—Hill } 8
" —School
" —Precinct—PO

Woods—School (2 of these)
" —Point
" —Island—Subd. } 6
" —Vill. —PO
" —Spring—Creek

Subd.—Lagoon
" — " —Subd.
" —Subd. } 5
" —Bay
" —School—Subd.

Point—Subd. (2 of these)
" —3 Subd. } 5
" —Landing
" —Bay

Prairie—School
" —Vill.—School
" — " —Town—PO } 5
" —PO —School
" — " ?—Station?—Vill.—School

Settl.—School
" —Vill.—School

School—Hill
" —PO—Vill.

Stat'n—School
" —2 Subd.

Camp—2 Subd.

County—PO—Town—Station—Vill.

Precinct—PO—Town—2 Vill.—School

Paper Town—PO—3 Vill.—PO—2 Schools—Subd.

As the above lists show, the feature from which Dane Co. names have most often spread to others is the Village or Corners name; next the PO name; next the Town name, and so on. The feature to which the name has most often spread first is the PO, next the village, and so on. Thus while the patterns are highly varied, some of them do dominate; the three largest groups begin from artificial-feature names, the three next largest from natural-feature names. More series begin from artificial than from natural (94 vs. 70) but even so the latter are proportionally more influential, since they are only half as numerous.

ANALYSIS OF THE CHARTS

The figures of the charts must be read with some caution. First, this list of names is surely not complete, and even if it were, the figures belong only to this particular area. A difference in general topography or in the degree or type of settlement would immediately be reflected in the place-names. At the same time, the writer has the strong impression that Dane Co. may prove a microcosm of the state of Wisconsin at large. It has considerable variation of topography, and in artificial features everything from the entirely rural to the urban. Great-lake, forest, and some other names are lacking which will figure in other counties, but there are samples of most other kinds in about typical distribution.

All names for the large features may be presumed to have been collected, but the smaller the features become the more likely they are to be overlooked (if they have names at all)—for example, countless springs, ponds, small creeks, crossroads. However, the figures on the charts permit some interesting observations:

1. The names for artificial features outnumber those for natural almost two to one, thanks to the large number of subdivisions, schools and PO's, which together furnish 60.04 percent.

2. Of natural-feature names, water-courses furnish 29.40 percent, relatively high lands 23.02 percent, bodies of water 17.79 percent; together, 70.21 percent.

3. In both natural and artificial, the most frequent types are the descriptive, and those for local people and for local places and institutions. These three account for over 83 percent of the names. The third is the largest among artificial-feature names because of the way names spread from one local feature to another, which occurs oftener in settlements.

CHART I

	NON-LOCAL PERSON	LOCAL PERSON	DISTANT PLACE	NEARBY PLACE	DESCR. AND LOC.	SUBJECTIVE DESCR.	INSPIR. SYMBOLIC	ANECDOTIC IRONIC	UNCERT. UNKNOWN	PUNNING BLENDING	PSEUDO-INDIAN	INDIAN	FRENCH	GERMAN	NORWEG'N	517
Spring		27		3	13						1		1		3	43
River				3	11				6			10	1			20
Creek		22	2	20	24			2				8				68
Branch (1)				1	1											2
Branch (2)				6	7											13
Brook					1											1
Run		1		1												2
Fork					1											1
Ditch					1											1
Canal					1											1
Lake		18		2	46	1			3			10				70
Pond		8		1	4											13
Spread					1											1
Slough		1			2											3
Lagoon				4	3								2			5
Harbor		1														1
Beach		4	1	1												6
Bay		3		5	2											10
Point (1)		10		3	11					1	1					25
Island		6			7	1										14
Marsh		5		3	2											10
Swamp		1			1											2
Prairie		2	1	8	4		1	1								16
Plain				1												1
Valley		11	2	5	4	3		1	1						4	25
Hollow		1		1	2			3								7
Hole					1											1
Pocket						1										1
Bottoms					2										1	2
Cave		4		2	1											7
Mountain					1											1
Mount		1			1	1										3
Mound					2		1									3
Hill	1	48		9	24	1	1	4	3					3		83
Ridge				5	2											7
Bluff		3		1	9											13
Hogback					2											2
Heights			1	1												2
Knoll	1															1
Point (2)					1				1							2
Bank					2											2
Rock		2		1	3	1										7
Stone					1											1
Terrace					1											1
Tree					2	1										3
Grove		2	1	1	2											6
Woods		6			2											8
575	2	184	8	86	208	10	3	11	14	1	2	28	4	3	8	
Percent	0.34	31.94	1.39	14.98	36.28	1.73	0.52	1.91	2.43	0.17	0.34	4.86	0.69	0.52	1.39	

CHART II

	NON-LOCAL PERSON	LOCAL PERSON	DISTANT PLACE	NEARBY PLACE	DESCR. AND LOC.	SUBJECTIVE DESCR.	INSPIR. SYMBOLIC	ANECDOTIC IRONIC	UNCERT. UNKNOWN	PUNNING BLENDING	PSEUDO-INDIAN	INDIAN	FRENCH	GERMAN	NORWEG'N	1036
County	1															1
Town	2	2	19	14	2	1		1	3							44
Div. of Area				6												6
City	1	1														2
Village	3	18	7	40	8	2		1	1	1		2	1			79
Subdivision	3	40	4	70	103	24	1		4	4	1	5	2			253
"Paper Town"	2	1	5	8		1			1							18
Settlement	1	12	2	2	7					1				1	1	24
Precinct				15												15
Post Office	3	10	8	83	4		4	2	6							120
Church		1	1	12	1		4									19
School		115		78	44	11	1									249
Tavern		5	1													6
House					3											3
Place		2														2
Stand		1														1
(Surname)'s		5														5
Corner(s)		33	1	31	8		1	1						1	1	75
Points					1											1
Crossroad(s)		2		1												3
Junction				1												1
Railroad				14												14
Highway				1												1
Road		2	2	23	3											30
Lane					3											3
Bend					1											1
Crossing				1	3											4
Ford				4	1							3				5
Bridge				1	1											2
Landing		6		1												7
Station		2		12	1			3								18
Camp	2			1	2											5
Field		1														1
Park		2			2									1		4
Forest						1										1
Farm		1	1		1	1	1			1					1	6
Breaking					1											1
Cut					1											1
Mine		1						1								2
Mill							1									1
Fort				1												1
Battleground																1
1057	16	263	51	420	202	41	13	9	15	7	1	10	3	3	3	
Percent	1.51	24.90	4.83	39.69	19.14	3.88	1.23	0.85	1.42	0.66	0.09	0.95	0.28	0.28	0.28	

4. Subjectively descriptive names are very few for natural features; they are found for several schools, and above all, subdivisions, which have as many as all other features combined.

5. Names for distant places and non-local people are from four to five times as frequent for artificial features as for natural.

6. Indian names, on the contrary, are five times as frequent for natural features as for artificial.

7. Punning and blending, anecdotic, ironic and humorous, inspirational and symbolic names, and those of non-Indian foreign languages are such small classes that the differences in the figures cannot be dependably interpreted.

APPENDIX II: Sources

CHIEF DOCUMENTARY SOURCES

(Except Standard Reference Works)

Abel	Abel, Henry I. *Geographical, Geological and Statistical Chart of Wisconsin and Iowa* Philadelphia, 1838
Adams	Adams, Bessie E. *History of Black Earth* W.H.S.MS., 1898
Ames	Ames, William L. *History of Oregon* [Oregon, Wis.], 1924
Argus	*Argus, The Daily* Madison, 1852
"	*Argus and Democrat, The Daily* Madison, 1852–3, 1860–2
"	*Argus and Democrat, The Weekly* Madison, 1852, 1855–8
"	*Argus, The Wisconsin* Madison, 1844–53
Barton	Barton, Albert O. "Dane County" in John G. Gregory, *Southwestern Wisconsin* Chicago, 1932
"	*The Old Spring Hotel* W.H.S.MS., 1917

Barton — *The Story of Primrose, 1831–95*
Madison

Belle. — *Belleville, Town of Montrose*
W.H.S.MS., n.d.

Bohn — Bohn, Mrs. Belle C.
Memories of Monona
W.H.S.MS., 1936

Bross. — Brossard, Eugene E., ed.
Wisconsin Annotations
Madison, 1930 (2d ed.)

Brown — Brown, Charles E.
The Arboretum Springs
W.H.S.MS., Box 7 (also other boxes, referred to by number)

" — "History of the Lake Mendota Region" in *Lake Mendota Origin and History*
Technical Club of Madison, 1936

" — "Lake Monona" in *The Wisconsin Archeologist* N.S. Vol. 1, No. 4

" — *Lake Mendota Indian Legends*
Madison, 1927

Butler — Butler, James Davie
"Taychoperah, the Four Lake Country" in W.H.S. *Collections*, X
Madison, 1885

Butt. — Butterfield, Consul W.
History of Dane County, Wisconsin
Chicago, 1880

Cas. 1 — Cassidy, Frederic G.
"The Naming of the Four Lakes" in *Wisconsin Magazine of History*, XXIX, 1
Madison, Sept. 1945

Cas. 2 — Cassidy, Frederic G.
"Koshkonong: A Misunderstood Place-Name" in *Wisconsin Magazine of History*, XXXI, 429
Madison, June 1948

Chap. — Chapman, C. B.
"Early Events in the Four Lake Country" in

W.H.S. *Collections*, IV
Madison, 1859

Childs
Childs, Col. Ebenezer
"Recollections of Wisconsin since 1820" in W.H.S. *Collections*, IV
Madison, 1859

Clark
Clark, Satterlee
"Early Times at Fort Winnebago" in W.H.S. *Collections*, VIII
Madison, 1879

Cole
Cole, Harry E.
Tavern Tales and Travel Trails
W.H.S.MS.

Dane
Dane County, W. T., Journal of the Board of Supervisors of, in Dane County Court House
1839 forward

Dela.
Delaplaine, George P.
"Geo. P. Delaplaine's Statement, Nov. 2, 1887" in W.H.S. *Collections*, XI
Madison, 1888

Demo.
Democrat, Wisconsin
Madison, 1842–44, 1846–51

"
Democrat, Wisconsin Daily
Madison, 1865–6

"
Democrat, Wisconsin Weekly
Madison, 1865–6

Doty L.
Doty, James D.
(Letter on the Orthography of Wisconsin Place-Names) in *The New York American*
Oct. 1, 1840

Draper
Draper, Lyman C.
Madison, the Capital of Wisconsin . . .
Madison, 1857

"
"Michel St. Cyr, an Early Dane County Pioneer" in W.H.S. *Collections*, VI
Madison, 1872

"
(Origin of the Naming of the Four Lakes) in *Draper Correspondence*
W.H.S.MS., (1877)

Dunk. — Johnson, Ella, and Erickson, Nora
Dunkirk
W.H.S.MS., n.d.

Durrie — Durrie, Daniel S.
A History of Madison, the Capital of Wisconsin . . .
Madison, 1874

Eldred — Eldred, Claude H.
Early Roads in the Vicinity of Sun Prairie
W.H.S.MS., Feb. 3, 1937

" — *Indian Trails in the Vicinity of Sun Prairie in 1837*
W.H.S.MS., Feb. 21, 1937

Expr. — *Express, Madison*
Madison, 1839–47

" — *Express, Wisconsin*
Madison, 1848–50, 1851–2

Feath. — Featherstonehaugh, George William
A Canoe Voyage up the Minnay Sotor . . .
London, 1847

" — *Report of a Geological Reconnoissance made in 1835 . . .*
Washington, 1836

For. — "Forgotten Cities of Dane County" in *Wis. State Journal*
Madison, Oct. 29, 1916

Foster — Foster, Mary, *et al.*
Wisconsin Historical Society File of Geographical Names
Madison, Historical Library

Hellum — Hellum, Olga
An Historic Sketch of the Township of Rutland
W.H.S.MS., n.d.

H. A. W. — *Historical Atlas of Wisconsin*
Snyder, Van Vechten & Co.
Milwaukee, 1878

Jones — Jones, C. E.
Madison: Its Origin, Institutions and Attractions
Madison, 1876

Keenan — Keenan, Harry A.
History of the Town of Dunn
W.H.S.MS., n.d.

Keyes 1 — Keyes, Elisha W., ed.
History of Dane County, Wisconsin
Madison, 1906

" 2 — "Early Days of Cambridge" in *The Cambridge News*
Cambridge, Wis., July 16, 1897

Kinzie — Kinzie, Mrs. John H.
Wau-Bun, the "Early Day" in the North-west
New York, 1856

Kittle — Kittle, William
History of the Township and Village of Mazomanie
Madison, 1900

Knapp — Knapp, J. G.
"Early Reminiscences of Madison" in W.H.S. *Collections*, VI
Madison, 1872

Kruse — Kruse, Louise
History of the Village of Middleton
W.H.S.MS., n.d.

Lemmon — Lemmon, J. D.
Dane County Almanac
Madison, 1872

Long — Long, Stephen H.
Voyage . . . to the Falls of Saint Anthony in 1817 in Minnesota H.S. *Collections*
Philadelphia, 1860

Lyon — Lyon, Orson
(*Note Book: Notes of Survey of the Madison Area*)
W.H.S.MS., 1834

M. C. T. — *Capital Times, The*
Madison, Dec. 13, 1917 ff., passim.

Marry. — Marryat, Capt. Frederick
Diary in America, pp. 185–205
Philadelphia, 1839 (also London and New York)

Martin	Martin, Morgan L. (*Letter, referring to his trip to the Four Lakes, 1829*) W.H.S.MS., 1885
"	"Narrative of Morgan L. Martin" in W.H.S. *Collections*, XI Madison, 1888
Mills	Mills, Simeon "A Short Chapter in the Early History of Dane County . . ." in The *Madison Democrat* Madison, Dec. 6 and 20, 1886
"	"The Four Lakes, How They Were Named" in Park, William J. *Madison, Dane County* . . . (see below)
Morse	Morse, the Rev. Jedidiah *Report to the Secretary of War of the United States on Indian Affairs* New Haven, 1822
Nak. Tom	Madison Realty Co. *Nakoma Tomahawk, The* Madison, 1920–1
Nilsen	Nilsen, Svein "De Skandinaviske Settlementer i Amerika" in *Billed-Magazin, I* Madison, 1869
Noonan	Noonan, Josiah A. "Recollections of Wisconsin in February 1837" in W.H.S. *Collections*, VII Madison, 1876
"	"Dane County" in *The Wisconsin Enquirer* May 25, 1839
Park	Park, William J. and Co. *Madison, Dane County and Surrounding Towns* . . . Madison, 1877
Peck	Peck, George Wilbur *Wisconsin, Comprising Sketches . . . in Cyclopedic Form*, Madison, 1906

Pedigo	Pedigo, J. Paul "The Great Cave of Dane County" in *The Madison Democrat* Madison, Dec. 29, 1918
Phil.	Phillips, Louise L. [*Scrapbook*, in possession of the W.H.S.] 1898
Pike	Pike, Lieut. *An Account of a Voyage up the Mississippi River . . . 1805-6* [Washington, 1807?]
Quam.	Quammen, John A. *Norwegian Pioneer Association, Biographical Record* W.H.S. MS., 1898
Regent	*Report of the Regents of the University of Wisconsin* Madison, 1849
Rogers	Rogers, Alfred T. [The Founding of Nakoma] in *The Nakoma Tomahawk* Madison, Apr., 1920
Rollis	Rollis, Capt. C. J. "Early History of Wheeler Prairie" in *The Madison Democrat* April 4, 1920
Ruste	Ruste, C. O. *Sixty Years of Perry Congregation* (Northfield, Minn., 1915)
Scha.	Schafer, Joseph *The Wisconsin Domesday Book* Wis. Hist. Soc. Addresses and Separates, No. 200 Madison, 1920
Schoo.	Schoolcraft, Henry R. *Information Respecting . . . The Indian Tribes of the U. S.* Philadelphia, 1851–7, 6 vols.

Seym.	Seymour, Wm. N. *Madison Directory and Business Advertiser* Madison, 1855
Sibley	Sibley, Henry Hastings *The Unfinished Autobiography of*—ed. Theodore C. Blegen Minneapolis, 1932
Smith I	Smith, Issac T. *Narrative of a trip in . . . 1838, to Pine River . . .,* W.H.S.MS.
Smith W	Smith, Gen. William R. *Observations on Wisconsin Territory* Philadelphia, 1838
Stat.	*Statistics of Dane County, Wisconsin* Madison, 1851, 1852
Steele	Steele, Robert, and Arries, Mansfield *History of the Town of Dane* in Park, Wm. J. *Madison, Dane County* . . . (see above)
Stew.	Stewart, George R. "The Source of the Name 'Oregon'" in *American Speech*, XIX, 2 Apr. 1944
Stoner	Stoner, Geo. W. "Stoner's Prairie" in *The Madison Democrat* May 15, 1904
Story	Story, H. E. "Montrose" in Park, Wm. J. *Madison, Dane County* . . . (see above)
Strong	Strong, Moses M. *History of the Territory of Wisconsin from 1836 to 1848* . . . Madison, 1885
Suckow	Suckow, B. W. *Madison City Directory* Madison, 1866
Surv.	Surveyors' Notebooks, 1833 and after MSS. in the Land Office, State Capitol Madison, Wis.

Tenney	Tenney, Major H. A. "Early Times in Wisconsin" (1849) in W.H.S. *Collections*, I Madison, 1854
Thw. 1	Thwaites, Reuben Gold "Early History of Belleville" in *The Belleville Recorder* Mar. 19, 1915
" 2	ed., *Early Western Travels* Cleveland, 1904–7, 32 vols.
" 3	"Names of the Lakes" in *Wisconsin State Jrnl.* Jan. 21, 1895
Treat.	*Treaties with the Winnebago Indians 1827–1865* Wis. Historical Library, Madison
Wakef.	Wakefield, John A. *History of the War between the United States and the Sac and Fox Nations of Indians* Jacksonville, Ill., 1834
Waun.	*History of Waunakee* W.H.S.MS., (1898)
West	West, Nathaniel, D. D. *The Ancestry, Life, and Times of Hon. Henry Hastings Sibley, LL.D.* St. Paul, 1889
Whit.	Whittet, Lawrence C. *Address delivered at Old Settlers' Picnic* Albion, Sept. 2, 1935
Whitt.	Whittlesey, Charles "Recollections of a Tour through Wisconsin in 1832" in W.H.S. *Collections*, I Madison, 1854
Will.	Willoughby, W. W. "Montrose" in Park, Wm. J. *Madison, Dane County* . . . (see above)
W.P.O.	*Wisconsin Historical Society File of Wisconsin Post Offices* Based on Federal Records Madison, Historical Library, 1918

W.S.J. *Wisconsin State Journal, The* Madison, Sept. 1852 ff., passim

LIST OF MAPS

Arranged Chronologically

(In possession of Wis. State Hist. Society)

Finley	1826	Finley, Anthony, *A New American Atlas*, Philadelphia.
Farmer	1828?	Farmer, John, *Sketch of the Fox and Wisconsin Rivers*, n.p.
Doty	1829	Doty, James D., *Route of Governor Doty from Green Bay to Prairie du Chien*, MS.
Chan.	1829	Chandler, R. W., *Map of the U. S. Lead Mines on the Upper Mississippi River*, Cincinnati.
Farmer	1830	Farmer, John, *The Surveyed Part of the Territory of Michigan*, Utica.
Farmer	1830	Farmer, John, *Map of the Territories of Michigan and Ouisconsin*, Albany.
Mitch.	1831	Mitchell, S. Augustus, *Map of the States of Ohio, Indiana and Illinois*, Philadelphia.
Center	1832	Center, A. J., *Map of the Route of the Military Road from Fort Crawford to Fort Howard, via Fort Winnebago*, War Dept., Office of Chief of Engin.
N. A.	1833	*The Northwest and Michigan Territories*, North America, sheet V.
Terr.	1835	*Map of the Surveyed Part of Wisconsin Territory*, Surveyor General's Office.
Feath.	1835	Featherstonehaugh, G. W., *A Portion of the Indian Country Lying East and West of the Mississippi River*, n.p.
Mitch.	1835	Mitchell, S. Augustus, *Map of the States of Ohio, Indiana and Illinois*, Philadelphia.
Burr	1836	Burr, David H., *Map of the Territory of Wisconsin*, n.p.

Suydam	1836	Suydam, John V., *Map of the Four Lake Country*, n.p.
Judson	1836	Judson, L., *Map of the Western Land District, Wisconsin*, N.Y.
Finley	1837	Finley, J. W., *Part of Wisconsin Including Madison. . .*, Baltimore.
Morr.	1837	Morrison, Samuel, *Topographical Map of the Wisconsin Territory*, Surveyor General's Office.
Burr	1838	Burr, David H., *Map of Illinois with Parts of Indiana, Wisconsin, etc.*, n.p.
Hinman	1838	Hinman and Dutton, *Map of the Settled Part of Wisconsin Territory*, Philadelphia.
I & W	1838	Bradford *Atlas: Iowa and Wisconsin*, n.p.
Judson	1838	Judson, L., *Map of the Entire Territories of Wisconsin and Iowa*, Cincinnati.
Suydam	1838	Suydam, J. V., *Map of Wisconsin Territory*, N. Y.
Taylor	1838	Taylor, Stephen, *Map of the Wisconsin Land District*, Philadelphia.
Cram	1839	Cram, Thos. J., *Map of Wiskonsin Territory*, Washington.
Hagner	1839	Hagner, Lieut. C. N., *Map of the Four Lakes*, Senate Doc. 318 (serial Vol. 359).
Tanner	1839	Tanner, Thos. R., *A New and Authentic Map of the State of Michigan and Territory of Wisconsin*, Philadelphia.
Doty	1844	Doty, Charles, and Hudson, Francis, *Map of Wiskonsan*, n.p.
Tanner	1845	Tanner, Thos. R., *A New Sectional Map of the State of Michigan and Territory of Wisconsin*, Philadelphia.
Jervis	1846	Jervis and Edgerton, *Wisconsin . . . Milwaukie Land District*, N.Y.

Lapham	1846	Lapham, I. A., *Wisconsin Sectional Map*, Milwaukee.
P. S.	1846	*A Sketch of the Public Surveys in Wisconsin*, MS. "From the I. A Lapham collection."
Morse	1847	Morse, S. E., and Breese, Samuel, *Wisconsin, Southern Part*, n.p.
Lapham	1848	Lapham, I. A., *Wisconsin Sectional Map*, Milwaukee.
"	1849	*The State of Wisconsin*, Milwaukee.
"	1850	*The State of Wisconsin*, Milwaukee.
"	1852	*The State of Wisconsin*, Milwaukee.
"	1853	*The State of Wisconsin*, Milwaukee.
Gree.	1854	Greeley, Horace, and Co., pub., *Map of Madison and the Four Lake Country*, N. Y.
"	1855	[Map of *Dane County*], N. Y.
Chap.	1855	Chapman, Silas, *Sectional Map of Wisconsin*, Milwaukee.
Lapham	1855	Lapham, I. A., *The State of Wisconsin*, Milwaukee.
Morse	1855	Morse, Chas. W., *Cerographic Map of Wisconsin*, Chicago.
Durrie	1856	Durrie, Daniel S., *Map of the City of Madison, 1856*, Madison.
Canf.	1859	Canfield, Wm. H., *Map of Sauk County Wisconsin*, Milwaukee.
Lig.	1861	Ligowski, A., *Map of Dane County, Wisconsin*, Madison.
H & W	1873	Harrison and Warner, *Atlas of Dane County Wisconsin*, Madison.
Warren	1875	Warren, G. K., [*Wisconsin River from Portage to Its Mouth*], n.p.

Foote	1890	Foote, C. M., and Henion, J. W., *Plat Book of Dane County Wisconsin*, Minneapolis.
Gay	1899	Gay, Leonard W., *New Atlas of Dane County Wisconsin*, Madison.
Mendo.	1900	Birge, E. A., *et al.*, *Hydrographic Map of Lake Mendota*, Madison.
Monona	1900	*Hydrographic Map of Lake Monona*, Madison.
Demo.	1904	Madison Daily Democrat, *Atlas of Dane County*, Madison.
Rural	1910	United States Post Office Dept., Rural Delivery Service, Dane Co.
Cant.	1911	Cantwell Printing Co., *Standard Historical Atlas of Dane County Wisconsin*, Madison.
Soils	1917	United States Dept. of Agriculture, Bureau of Soils, Soils Map, Wisconsin, Dane Co. Sheet.
Brown	1924	Brown, Charles E., dir., *Wisconsin Archeological Atlas*, in Wisconsin Historical Library, Madison.
Blied	1926	Blied Printing Co., *New Atlas of Dane County*, Madison.
Thrift	1931	Thrift Press, The, *Atlas and Plat Book of Dane County Wisconsin*, Rockford.
Hixson	1935	Hixson, W. W., *Dane County Plat Book*, Rockford.
USGS	v.d.	*United States Geological Survey Topographic Maps:*
	1887–9	Stoughton Sheet
	1892	Baraboo Sheet
	1899	Poynette Quadrangle
	1904	Evansville Quadrangle
	1904	Madison Quadrangle (ed. of 1892 not available)

	1905	Sun Prairie Quadrangle
	1906–7	Cross Plains Quadrangle
	1916–20	Blue Mounds Quadrangle
	1919–20	New Glarus Quadrangle
	1920–1	Blanchardville Quadrangle
Hiway	1940, etc.	State Highway Commission, *County Maps of Wisconsin*, Madison.

LIST OF INFORMANTS

Abbr.	*Name*	*Residence*
Adler	Adler, Michael	Ashton Corners
Ames	Ames, W. L.	Oregon
Ander.	Anderson, Amos	Town of Vermont
Anth.	Anthony, David C.	" " Rutland
Austin	Austin, Henry	Mt. Horeb
Bab.	Babcock, Milton J.	Town of Albion
Barry	Barry, George	" " Fitchburg
Barsn.	Barsness, Mrs. Ed.	Black Earth
*Barton	Barton, Albert O.	Madison
Bent.	Bentley, Wayne	Madison
Betl.	Betlach, Mrs. Mary	Sun Prairie
Boning	Boning, Miss Margaret	Town of Montrose
Brere.	Brereton, Lawson	" " Dane
Briggs	Briggs, Mrs. Russell	Madison
*Brig.	Brigham, Charles I.	Blue Mounds
*Brown	Brown, Charles E.	Madison
Butz	Butz, Mrs. Julius	Mazomanie
*Cass	Cass, Betty	Madison
Clapp	Clapp, Hector	Madison
Clough	Clough, Proctor	Town of Albion
Coll.	Colladay, Charles W.	" " Dunn
Cong.	Congdon, Mrs.	Oregon
*Corc.	Corcoran, Richard	Town of Westport
Corsc.	Corscot, Garrett	Madison
Cunn.	Cunningham, F. J.	Cottage Grove
Dahlk	Dahlk, Elmer	Town of Middleton
*Dahmen	Dahmen, A.	Cross Plains

* = Informants who gave particularly useful information.

Davids.	Davidson, Mrs. Adam and daughters	Verona
DeBow.	DeBower, Edward	Dane
Deneen	Deneen, Miss Mamie	Town of Vermont
*Derl.	Derleth, August	Sauk City
Dow	Dow, Robe, Jr.	Stoughton
Dusch.	Duscheck, Mr. and Mrs. Edward	Sun Prairie
Dybd.	Dybdahl, Miss Viola	Black Earth
Edw.	Edwards, Theodore	McFarland
Ellis	Ellis, C. F.	Windsor
Espes.	Espeseth, Mrs. Mabel	Mt. Horeb
Femr.	Femrite, H. O.	Madison
Flint	Flint, A. T.	Madison
Fraut.	Frautschi, Walter	Madison
Frey	Frey, N. J.	Madison
Fritz	Fritz, Edward	Belleville
Gal.	Gallistel, Mrs. A. F.	Madison
Gannon	Gannon, William	Oregon
*Geary	Geary, the Rev. James A.	Washington, D. C.
*Gilb.	Gilbertson, Gilbert	Mt. Horeb
Godd.	Goddard, Mrs. Elmer J.	Town of Dane
Goebel	Goebel, Frank	Town of Perry
Good	Good, Mr. and Mrs. Frank	Town of Cottage Grove
*Haugen	Haugen, Prof. Einar	Madison
Henn.	Henning, Miss Wilma	Sauk City
Henry	Henry, Mrs. E. J.	Basco
Horsw.	Horswill, Mrs. W. K.	Mazomanie
Johns.	Johnson, Mr. and Mrs. Julius	Stoughton
Juve	Juve, Sievert	Town of Pleasant Sprs.
Keenan	Keenan, Mr. and Mrs. Wm.	Town of Dunn
Kenn.	Kennedy, Wm. and Forrest	Town of Cottage Grove
Ketelb.	Ketelboeter, Harold	Town of Berry
Kilian	Kilian, Thos. F.	Town of Cottage Grove

Kitls.	Kittelson, William	Blanchardville
Kund.	Kundert, Mrs. Dorothy	Monroe
Lake	Lake, Charles	Windsor
Lawes	Lawes, Miss Genie	Mazomanie
Lein	Lein, Lars O.	Town of Albion
Lien	Lien, Mrs. Alyce	Blanchardville
Linc.	Lincoln, Frank	Black Earth
Lincn.	Lincoln, Harry	Tama, Iowa
Linde	Linde, Andrew	DeForest
Long	Long, Miss Pearl	Oregon
Loper	Loper, Mrs. Mary	Lodi
Mack	Mack, Melchior	Roxbury
MacK.	MacKenzie, John B.	North Lake
Maher	Maher, Mrs. Nina	Madison
Mars.	Marsden, Leonard	Town of Albion
McFar.	McFarland, Joe	McFarland
McKen.	McKenna, J. C., Sr. and Jr.	Madison
Mick.	Mickelson, Elmer	Town of Vermont
Minch	Minch, Karl	Belleville
Moore	Moore, F. G.	Mt. Vernon
Mor.	Morrison, Mrs. Edith	Morrisonville
Mussen	Mussen, Frank	Town of Roxbury
Newt.	Newton, Frank	Town of Rutland
Olson	Olson, Jim	Town of Cottage Grove
Oncken	Oncken, George	Waunakee
Onstad	Onstad, Otto	Madison
Ottes.	Otteson, S. A.	Oregon
Owen	Owen, Ray S.	Madison
Pauls.	Paulson, Mrs. Mattie	Mt. Horeb
Pets.	Peterson, Peter S.	Town of Rutland
Pfaff	Pfaff, Frank	Sun Prairie
Pick.	Pickering, Mrs. A. W.	Black Earth
Purc.	Purcell, Mrs. John	Town of Fitchburg
Quam	Quam, Lars O., and mother	Town of Dunn
Renk	Renk, William	Town of Bristol
*Reque	Reque, Mrs. Stark	Madison

Rowley	Rowley, Dr. A. G.	Middleton
Schor.	Schorger, A. W.	Madison
Schul.	Schultz, Wesley C.	Mazomanie
Simon	Simon, Mrs. Ferne	Waunakee
Skjo.	Skjolass, Mrs. G.	Town of Dunkirk
Smith	Smith, Alice E.	Madison
Soren.	Sorenson, E. P.	Marshall
Spaan.	Spaanem, Thore	Mt. Horeb
Stark.	Starkweather, F. W.	Madison
Steele	Steele, Mr. and Mrs. Ed.	Lodi
Stew.	Stewart, Frank	Verona
*Stucki	Stucki, The Rev. Benjamin	Neilsville
Sull.	Sullivan, Dr. A. G.	Madison
Sutter	Sutter, Louis	Madison
Swig.	Swiggum, Sam	Belleville
*Thor.	Thorstad, Jens	Town of Deerfield
Toban	Toban, Mrs. L. H.	Town of Westport
Utter	Utter, Jesse	Town of Rutland
Valen.	Valentine, M. P.	Cross Plains
Vroman	Vroman, Elmer	Town of Fitchburg
Walser	Walser, Peter	Marxville
Weum	Weum, Lefferd	Cambridge
Wied.	Wiedenbeck, F. J.	New London
Wilt	Wilt, Albert	Marshall
Wood	Wood, Byron	Burke
*Woodw.	Woodward, W. L.	Madison
Yelk	Yelk, Walter	Marshall